I0019926

FIBER OPTICS

NETWORKING AND DATA TRANSMISSION IN ACTION

4 BOOKS IN 1

BOOK 1
FIBER OPTICS 101: A BEGINNER'S GUIDE TO NETWORKING AND DATA
TRANSMISSION

BOOK 2
MASTERING FIBER OPTIC NETWORKS: ADVANCED TECHNIQUES AND
APPLICATIONS

BOOK 3
FIBER OPTIC INFRASTRUCTURE DESIGN AND IMPLEMENTATION:
PRACTICAL STRATEGIES FOR PROFESSIONALS

BOOK 4
CUTTING-EDGE FIBER OPTICS: EMERGING TECHNOLOGIES AND FUTURE
TRENDS IN NETWORKING

ROB BOTWRIGHT

Published by Rob Botwright
Library of Congress Cataloging-in-Publication Data
ISBN 978-1-83938-713-5
Cover design by Rizzo

Disclaimer

The contents of this book are based on extensive research and the best available historical sources. However, the author and publisher make no claims, promises, or guarantees about the accuracy, completeness, or adequacy of the information contained herein. The information in this book is provided on an "as is" basis, and the author and publisher disclaim any and all liability for any errors, omissions, or inaccuracies in the information or for any actions taken in reliance on such information. The opinions and views expressed in this book are those of the author and do not necessarily reflect the official policy or position of any organization or individual mentioned in this book. Any reference to specific people, places, or events is intended only to provide historical context and is not intended to defame or malign any group, individual, or entity. The information in this book is intended for educational and entertainment purposes only. It is not intended to be a substitute for professional advice or judgment. Readers are encouraged to conduct their own research and to seek professional advice where appropriate. Every effort has been made to obtain necessary permissions and acknowledgments for all images and other copyrighted material used in this book. Any errors or omissions in this regard are unintentional, and the author and publisher will correct them in future editions.

BOOK 1 - FIBER OPTICS 101: A BEGINNER'S GUIDE TO NETWORKING AND DATA TRANSMISSION

BOOK 2 - MASTERING FIBER OPTIC NETWORKS: ADVANCED TECHNIQUES AND APPLICATIONS

BOOK 3 - FIBER OPTIC INFRASTRUCTURE DESIGN AND IMPLEMENTATION: PRACTICAL STRATEGIES FOR PROFESSIONALS

BOOK 4 - CUTTING-EDGE FIBER OPTICS: EMERGING TECHNOLOGIES AND FUTURE TRENDS IN NETWORKING

Introduction

Welcome to the world of fiber optics, where the speed of light powers our modern communication networks and data transmission systems. The "Fiber Optics: Networking and Data Transmission in Action" book bundle offers a comprehensive exploration of this cutting-edge technology, providing readers with a deep understanding of its principles, applications, and future possibilities.

Book 1, "Fiber Optics 101: A Beginner's Guide to Networking and Data Transmission," serves as the perfect introduction for those new to the field. It covers the fundamental concepts of fiber optics, including how light travels through optical fibers, the basics of networking protocols, and the principles behind data transmission. Whether you're a student, an aspiring technician, or simply curious about this fascinating technology, this book will provide you with a solid foundation to build upon.

Building upon the foundational knowledge provided in Book 1, Book 2, "Mastering Fiber Optic Networks: Advanced Techniques and Applications," delves into the intricacies of fiber optic networking. From advanced techniques for signal modulation and multiplexing to practical applications in telecommunications, healthcare, and beyond, this book equips readers with the skills and insights needed to optimize fiber optic networks for various scenarios.

For professionals involved in the design and implementation of fiber optic infrastructure, Book 3, "Fiber Optic Infrastructure Design and Implementation: Practical Strategies for Professionals," offers invaluable guidance. This book covers everything from planning and deployment strategies to troubleshooting techniques, providing practical advice and real-world examples to help professionals navigate the complexities of fiber optic projects successfully.

Finally, Book 4, "Cutting-Edge Fiber Optics: Emerging Technologies and Future Trends in Networking," explores the latest advancements and future trends shaping the field of fiber optics. From quantum communication and terahertz transmission to photonic integrated circuits and beyond, this book offers a glimpse into the exciting innovations that will drive the next generation of fiber optic networks.

Whether you're a novice seeking to learn the basics, an experienced professional looking to expand your knowledge, or simply curious about the future of networking technology, the "Fiber Optics" book bundle has something for everyone. Join us on a journey through the fascinating world of fiber optics, where the possibilities are as limitless as the speed of light itself.

BOOK 1
FIBER OPTICS 101
A BEGINNER'S GUIDE TO NETWORKING AND DATA
TRANSMISSION

ROB BOTWRIGHT

Chapter 1: Introduction to Fiber Optics

Fiber optics, a technology that revolutionized communication systems, traces its roots back to the early 19th century. The concept of transmitting light through transparent materials was first explored by Daniel Colladon and Jacques Babinet in the 1840s. However, it wasn't until the 20th century that significant advancements in fiber optic technology occurred. In the 1950s and 1960s, researchers began experimenting with glass fibers for transmitting light, laying the groundwork for modern fiber optic communication.

The breakthrough came in the 1970s when Corning Glass Works introduced low-loss optical fibers made of fused silica. These fibers drastically reduced signal loss, making long-distance communication feasible. Around the same time, developments in semiconductor technology led to the invention of the semiconductor laser, which became the light source for fiber optic systems. These advancements paved the way for the deployment of fiber optic cables in telecommunications networks.

Throughout the 1980s and 1990s, fiber optic communication networks rapidly expanded, driven by the increasing demand for high-speed data transmission. Fiber optic cables offered unparalleled bandwidth and reliability compared to traditional copper cables. As a result, they became the backbone of global telecommunications infrastructure, supporting internet, telephone, and television services.

The early 21st century witnessed further innovations in fiber optic technology. The introduction of wavelength-division multiplexing (WDM) allowed multiple data streams to be transmitted simultaneously over a single fiber, significantly increasing capacity. This, coupled with the deployment of optical amplifiers, enabled the construction of ultra-long-haul fiber optic networks spanning thousands of kilometers without the need for signal regeneration.

Moreover, advancements in fiber optic sensing opened up new possibilities in various industries, including healthcare, aerospace, and oil and gas. Fiber optic sensors, capable of measuring parameters such as temperature, pressure, and strain with high accuracy and reliability, found applications in structural health monitoring, environmental monitoring, and industrial process control.

As fiber optic technology continued to evolve, researchers explored new materials and techniques to further enhance performance and functionality. One such innovation is photonic crystal fibers (PCFs), which feature a periodic microstructured cladding that enables unique light guidance properties. PCFs offer unprecedented control over light propagation, enabling applications in areas such as nonlinear optics, supercontinuum generation, and high-power laser delivery.

Furthermore, the integration of artificial intelligence (AI) and machine learning (ML) into fiber optic networks has emerged as a promising trend in recent years. AI algorithms can analyze vast amounts of network data in real-time, allowing operators to optimize performance, detect anomalies, and predict potential failures. ML techniques, coupled with advanced monitoring systems, enable proactive maintenance and fault prediction, minimizing downtime and improving overall network reliability.

Looking ahead, the future of fiber optics holds exciting possibilities. Emerging technologies such as quantum communication and terahertz communication are poised to revolutionize the field once again. Quantum key distribution (QKD) offers unparalleled security for data transmission by leveraging the principles of quantum mechanics to encrypt messages with unbreakable codes. Terahertz communication systems, operating at frequencies higher than traditional microwave systems, promise ultra-fast data rates and new applications in wireless networking and imaging.

In summary, the history and evolution of fiber optics have been characterized by continuous innovation and technological breakthroughs. From humble beginnings as an experimental curiosity to becoming the backbone of modern telecommunications, fiber optic technology has transformed the way we communicate, work, and live. As we look towards the future, the journey of fiber optics continues, driven by the quest for greater speed, capacity, and reliability in communication networks.

Fiber optics, with its exceptional properties and versatility, has found wide-ranging applications across various sectors of modern technology. From telecommunications to healthcare, from aerospace to automotive industries, fiber optic technology has become indispensable in enabling numerous cutting-edge applications.

In the realm of telecommunications, fiber optics reigns supreme as the backbone of global communication networks. The high bandwidth and low signal attenuation of optical fibers make them ideal for transmitting vast amounts of data over long distances with minimal loss. Internet service providers (ISPs) rely heavily on fiber optic cables to deliver high-speed internet access to homes and businesses. In deploying fiber optic networks, CLI commands such as "fiber-optic-cable-install" are commonly used to configure and manage network infrastructure. Additionally, wavelength-division multiplexing (WDM) techniques allow multiple data streams to be transmitted simultaneously over a single fiber, maximizing network capacity and efficiency.

Moreover, fiber optics has revolutionized the field of medicine, enabling advanced diagnostic and therapeutic techniques. In minimally invasive surgeries, fiber optic endoscopes equipped with miniature cameras and light sources provide surgeons with real-time visualization inside the body, allowing for precise interventions with minimal trauma to patients. Optical coherence tomography (OCT), another application of fiber optics, enables high-resolution imaging of biological tissues, aiding in the early detection and treatment of diseases such as cancer and retinal disorders. Deploying fiber optic sensors in healthcare settings requires meticulous calibration and configuration, often achieved through specialized software interfaces.

The aerospace industry has also embraced fiber optic technology for its lightweight and high-performance characteristics. In aircraft, fiber optic sensors are employed for structural health monitoring, detecting and assessing damage or stress in critical components such as wings, fuselage, and landing gear. These sensors can withstand extreme temperatures and harsh environments, making them ideal for aerospace applications where reliability is paramount. CLI commands such as "fiber-optic-sensor-deploy" are used to install and integrate fiber optic sensor systems into aircraft systems, ensuring seamless operation and accurate data collection.

Furthermore, the automotive sector has seen increasing adoption of fiber optic technology in vehicle design and manufacturing. Fiber optic cables are used for high-speed data transmission within automotive electronics systems, supporting features such as infotainment systems, GPS navigation, and advanced driver assistance systems (ADAS). By replacing traditional copper wiring with fiber optics, automotive manufacturers can reduce weight, improve fuel efficiency, and enhance overall vehicle performance. Deploying fiber optic networks in automotive applications involves specialized connectors and termination techniques to ensure reliable connectivity and signal integrity in the harsh conditions of automotive environments.

In the field of industrial automation and control, fiber optics plays a crucial role in enabling real-time monitoring and control of manufacturing processes. Fiber optic sensors are employed for measuring parameters such as temperature, pressure, and vibration in industrial machinery, ensuring optimal performance and preventing costly downtime. Additionally, fiber optic communication networks facilitate data exchange between sensors, actuators, and control systems, enabling seamless integration and automation of manufacturing processes. CLI commands such as "fiber-optic-network-config" are used to set up and configure fiber optic communication links in industrial environments, allowing for efficient data transmission and control.

Moreover, fiber optic technology has opened up new possibilities in the field of renewable energy, particularly in the monitoring and management of solar and wind power systems. Fiber optic sensors are utilized for monitoring temperature, strain, and wind speed in solar panels and wind turbines, optimizing energy production and ensuring operational safety. By integrating fiber optic sensors into renewable energy infrastructure, operators can detect potential faults or performance issues early on, minimizing downtime and maximizing energy output. Deploying fiber optic sensors in renewable energy applications requires careful calibration and alignment to ensure accurate and reliable measurement of environmental parameters.

In summary, fiber optics has become an indispensable tool in modern technology, driving innovation and advancement across various industries. From telecommunications to healthcare, from aerospace to automotive sectors, fiber optic technology continues to revolutionize the way we communicate, work, and live. As we look to the future, the applications of fiber optics are poised to expand further, unlocking new possibilities and shaping the technological landscape for generations to come.

Chapter 2: Basics of Light and Optics

Understanding the properties of light is fundamental to comprehending the behavior of electromagnetic radiation and its interaction with matter. Light, as a form of electromagnetic wave, exhibits various characteristics that define its nature and influence its applications across different fields of science and technology. One of the fundamental properties of light is its dual nature, manifesting as both waves and particles. This duality, described by the wave-particle duality principle, means that light can behave as both a wave and a stream of particles called photons. The wave nature of light is characterized by its frequency, wavelength, and amplitude, while its particle nature is associated with discrete packets of energy known as photons. CLI commands such as "light-properties-analysis" can be used to analyze the characteristics of light waves and photons, providing insights into their behavior and interactions in various optical systems.

Moreover, the speed of light is a fundamental property that plays a crucial role in many aspects of physics and engineering. In a vacuum, light travels at a constant speed of approximately 299,792 kilometers per second (or about 186,282 miles per second), often denoted by the symbol c. This universal constant serves as a foundational principle in the theory of relativity and forms the basis for defining units of length and time. In practical applications, the speed of light is a critical parameter in optical communications, where it determines the transmission latency and bandwidth of fiber optic networks. CLI commands such as "light-speed-calibrate" are used to measure and calibrate the speed of light in optical systems, ensuring accurate data transmission and synchronization.

Furthermore, light exhibits the phenomena of reflection, refraction, diffraction, and interference, which are essential for understanding how light interacts with surfaces and optical elements. Reflection occurs when light bounces off a surface, following the law of reflection that states that the angle of incidence is equal to the angle of reflection. Refraction, on the other hand, occurs when light passes from one medium to another, causing it to change direction due to differences in the refractive indices of the materials involved. Snell's law governs the relationship between the angles of incidence and refraction in refractive media. CLI commands such as "light-reflection-simulate" and "light-refraction-calculate" can be used to simulate and calculate the behavior of light at reflective and refractive interfaces, aiding in the design and optimization of optical systems.

Moreover, diffraction refers to the bending of light waves around obstacles or through apertures, resulting in the spreading of light beyond the geometric shadow of the obstacle. Diffraction patterns, characterized by alternating bright and dark fringes, are observed when light encounters obstacles with dimensions comparable to its wavelength. Interference, another fundamental property of light, occurs when two or more light waves overlap and either reinforce (constructive interference) or cancel out (destructive interference) each other. These phenomena are exploited in various optical devices and techniques, such as diffraction gratings and interferometers, for applications ranging from spectroscopy to laser beam shaping. CLI commands such as "light-diffraction-simulate" and "light-interference-analysis" enable engineers and scientists to model and analyze diffraction and interference effects in optical systems, facilitating the design and optimization of devices and experiments.

Additionally, the polarization of light is a property that describes the orientation of its electric field vector as it propagates through space. Light waves can be linearly polarized, circularly polarized, or elliptically polarized, depending on the orientation and magnitude of the electric field vector. Polarization plays a crucial role in many optical technologies, such as liquid crystal displays (LCDs), polarizing filters, and optical microscopy. CLI commands such as "light-polarization-calibrate" are used to control and manipulate the polarization state of light in optical systems, enabling polarization-sensitive applications and experiments.

Furthermore, light interacts with matter through absorption, transmission, and emission processes, leading to various optical phenomena observed in materials. Absorption occurs when light is absorbed by atoms or molecules, causing them to transition to higher energy states. Transmission refers to the passage of light through a material without significant attenuation or absorption, depending on the material's transparency and optical properties. Emission occurs when excited atoms or molecules return to lower energy states, releasing photons in the process. These processes are exploited in spectroscopy, fluorescence microscopy, and laser technology for applications ranging from chemical analysis to biomedical imaging. CLI commands such as "light-absorption-spectrum" and "light-emission-simulation" allow researchers and engineers to analyze and simulate the interactions between light and matter, facilitating the development of advanced optical devices and techniques.

In summary, the properties of light are fundamental to our understanding of the behavior of electromagnetic radiation and its applications in various scientific and technological fields. From its wave-particle duality to its speed, reflection, refraction, diffraction, interference, polarization, and interactions with matter, light exhibits a rich array of characteristics that underpin its myriad applications. By leveraging our knowledge of light's properties and employing advanced tools and techniques, we can continue to unlock new discoveries and innovations in optics, photonics, and beyond.

The laws of reflection and refraction govern the behavior of light as it interacts with surfaces and transitions between different media, providing fundamental principles that underpin many optical phenomena and devices. Reflection, the process by which light bounces off a surface, follows two primary laws: the law of reflection and the law of specular reflection. According to the law of reflection, the angle of incidence, denoted by θi, is equal to the angle of reflection, denoted by θr, measured from the normal to the surface. This geometric relationship holds true for all types of reflective surfaces, including mirrors, smooth surfaces, and interfaces between media. CLI commands such as "reflection-law-calculate" can be used to calculate the angle of reflection based on the angle of incidence, facilitating the design and analysis of optical systems.

Moreover, the law of specular reflection specifies that the reflected light rays are parallel to each other, forming a coherent beam of reflected light. This characteristic property of specular reflection is exploited in various optical devices, such as mirrors, lenses, and retroreflectors, for applications ranging from imaging and illumination to laser beam steering and signal transmission. CLI commands such as "specular-reflection-simulate" enable engineers and scientists to simulate and visualize the behavior of specular reflection in optical systems, aiding in the design and optimization of devices and experiments.

Additionally, refraction, the bending of light as it passes from one medium to another, is governed by Snell's law, named after the Dutch scientist Willebrord Snellius. According to Snell's law, the ratio of the sine of the angle of incidence to the sine of the angle of refraction is equal to the ratio of the refractive indices of the two media. Mathematically, Snell's law is expressed as $n_1 sin(\vartheta_i) = n_2 sin(\theta_r)$, where n_1 and n_2 are the refractive indices of the incident and refractive media, respectively, and θ_i and θ_r are the angles of incidence and refraction, measured with respect to the normal to the interface. CLI commands such as "refraction-law-calculate" can be used to calculate the angle of refraction based on the refractive indices of the incident and refractive media, facilitating the design and analysis of optical systems involving refraction.

Furthermore, the phenomenon of total internal reflection occurs when light traveling in a medium with a higher refractive index encounters an interface with a medium of lower refractive index at an angle of incidence greater than the critical angle. In this scenario, all of the incident light is reflected back into the higher refractive index medium, with no transmission into the lower refractive index medium. Total internal reflection is exploited in optical fibers, prism-based imaging systems, and fiber optic sensors for applications such as data transmission, spectroscopy, and sensing. CLI commands such as "total-internal-reflection-simulate" enable engineers and scientists to simulate and analyze total internal reflection phenomena in optical systems, aiding in the design and optimization of devices and experiments.

Moreover, the concept of optical dispersion, wherein different wavelengths of light experience different degrees of refraction as they pass through a refractive medium, is central to the formation of rainbows and the operation of optical prisms and lenses. Dispersion causes the separation of white light into its constituent colors, forming a spectrum ranging from red to violet. This phenomenon is exploited in spectroscopy and optical communications for wavelength-dependent signal processing and analysis. CLI commands such as "optical-dispersion-calculate" can be used to calculate the dispersion angle and wavelength-dependent refractive index of a given medium, facilitating the design and optimization of optical systems involving dispersion.

In summary, the laws of reflection and refraction provide fundamental principles that govern the behavior of light as it interacts with surfaces and transitions between different media. From the law of reflection, which describes the geometric relationship between incident and reflected light rays, to Snell's law, which quantifies the bending of light at optical interfaces, these laws underpin many optical phenomena and devices essential for various scientific, engineering, and technological applications. By understanding and leveraging these laws, engineers and scientists can design and develop innovative optical systems and devices that shape our modern world.

Chapter 3: Understanding Fiber Optic Cables

Fiber optic cables come in various types, each designed to meet specific performance requirements and deployment needs in optical communication systems. One of the most common types is single-mode fiber optic cable, characterized by a small core diameter (typically around 9 micrometers) and a single transmission mode. Single-mode fiber is well-suited for long-distance transmission of optical signals due to its low dispersion and attenuation characteristics. CLI commands such as "single-mode-fiber-deploy" can be used to deploy single-mode fiber optic cables in long-haul telecommunications networks, ensuring high-speed and reliable data transmission over extended distances.

In contrast, multi-mode fiber optic cable features a larger core diameter (typically around 50 or 62.5 micrometers) and supports multiple transmission modes. Multi-mode fiber is commonly used for short-distance communication, such as within buildings or data centers, where high bandwidth and cost-effectiveness are primary considerations. CLI commands such as "multi-mode-fiber-install" are used to install multi-mode fiber optic cables in premises cabling systems, providing high-speed connectivity for local area networks (LANs) and storage area networks (SANs). Furthermore, there are specialty fiber optic cables designed for specific applications and environments. One example is dispersion-shifted fiber, which is optimized to minimize chromatic dispersion, a phenomenon that causes different wavelengths of light to travel at different speeds and results in signal distortion. Dispersion-shifted fiber is commonly used in long-haul telecommunications networks where minimizing dispersion is critical for maintaining signal integrity over extended distances. CLI commands such as "dispersion-shifted-fiber-configure" can be used to configure dispersion-shifted fiber optic cables in high-speed backbone networks, ensuring optimal performance and reliability.

Another specialty fiber optic cable is polarization-maintaining fiber, which is engineered to maintain the polarization state of light as it propagates through the fiber. Polarization-maintaining fiber is used in applications such as fiber optic gyroscopes, interferometric sensors, and coherent optical communication systems, where preserving the polarization state of light is essential for accurate measurement and signal detection. CLI commands such as "polarization-maintaining-fiber-deploy" are used to deploy polarization-maintaining fiber optic cables in precision sensing and measurement applications, ensuring precise and reliable operation in demanding environments.

Moreover, there are armored fiber optic cables designed to withstand harsh environmental conditions and physical damage. Armored cables feature an additional layer of protection, such as metal or Kevlar reinforcement, to shield the fiber optic core from moisture, chemicals, abrasion, and impact. Armored fiber optic cables are commonly used in outdoor and industrial applications, such as military communications, oil and gas exploration, and industrial automation, where ruggedness and durability are paramount. CLI commands such as "armored-fiber-optic-cable-install" are used to install armored fiber optic cables in outdoor environments and industrial facilities, providing robust and reliable connectivity in challenging conditions.

Furthermore, hybrid fiber optic cables combine optical fibers with electrical conductors, such as copper wires or coaxial cables, to provide both optical and electrical connectivity in a single cable assembly. Hybrid cables are used in applications such as fiber-to-the-home (FTTH) networks, where optical fibers are used for high-speed data transmission, and copper wires are used for power delivery or telephone services. CLI commands such as "hybrid-fiber-optic-cable-configure" are used to configure hybrid fiber optic cables in FTTH deployments, ensuring seamless integration and operation of optical and electrical components.

Additionally, there are bend-insensitive fiber optic cables designed to minimize signal loss and attenuation when bent or twisted, making them ideal for installation in tight spaces or around corners. Bend-insensitive fiber optic cables feature a special fiber design or protective coating that reduces the impact of bending-induced losses, allowing for greater flexibility and ease of installation. These cables are used in applications such as fiber-to-the-desk (FTTD) installations, where space constraints and flexibility are important considerations. CLI commands such as "bend-insensitive-fiber-optic-cable-deploy" are used to deploy bend-insensitive fiber optic cables in office environments and confined spaces, ensuring reliable connectivity without sacrificing performance.

In summary, the various types of fiber optic cables offer a range of options to meet the diverse requirements of optical communication systems and applications. From single-mode and multi-mode fibers to dispersion-shifted, polarization-maintaining, armored, hybrid, and bend-insensitive cables, each type is designed to provide specific performance characteristics and functionality to address different deployment scenarios and environmental conditions. By understanding the characteristics and applications of each type of fiber optic cable, engineers and network operators can make informed decisions when designing and deploying optical communication networks.

Understanding the construction and characteristics of fiber optic cables is essential for designing and deploying reliable and efficient optical communication networks. Fiber optic cables consist of multiple components, each serving a specific purpose in protecting and transmitting optical signals over long distances. The primary components of a fiber optic cable include the core, cladding, buffer, strength member, and outer jacket. The core is the central region of the fiber optic cable through which light is transmitted, while the cladding is a layer of lower refractive index material surrounding the core, designed to confine the light within the core through total internal reflection. CLI commands such as "fiber-optic-cable-construction-visualize" can be used to visualize the construction of fiber optic cables, providing insights into the arrangement and composition of the core, cladding, and other components.

Moreover, the buffer serves as a protective layer around the optical fiber, providing cushioning and insulation against mechanical stress, moisture, and environmental factors. Buffers can be made of materials such as acrylate, silicone, or polyethylene, depending on the application and deployment environment. Strength members, typically made of aramid fibers (e.g., Kevlar) or fiberglass, provide additional mechanical support and tensile strength to the cable, preventing stretching or breakage during installation and operation. CLI commands such as "fiber-optic-cable-strength-test" are used to perform strength tests on fiber optic cables, ensuring they meet specified tensile and bend radius requirements for installation in various environments.

Furthermore, the outer jacket of a fiber optic cable serves as the primary protective layer, shielding the internal components from physical damage, moisture, chemicals, and UV radiation. Outer jackets are typically made of materials such as polyvinyl chloride (PVC), polyethylene (PE), or polyurethane (PU), chosen for their durability, flexibility, and resistance to environmental hazards. In outdoor or harsh environments, armored jackets may be used to provide additional protection against abrasion, impact, and rodent damage. CLI commands such as "fiber-optic-cable-jacket-inspect" are used to inspect the outer jacket of fiber optic cables for signs of damage or degradation, ensuring the integrity and longevity of the cable in service.

Additionally, fiber optic cables are classified based on their construction and performance characteristics, including mode of propagation, fiber count, and environmental rating. The mode of propagation refers to the number of transmission modes supported by the fiber, with single-mode and multi-mode being the two primary categories. Single-mode fibers have a small core diameter and support a single transmission mode, making them suitable for long-distance transmission with low dispersion and attenuation. Multi-mode fibers have a larger core diameter and support multiple transmission modes, making them suitable for short-distance transmission within buildings or data centers. CLI commands such as "fiber-optic-cable-mode-check" can be used to check the mode of propagation of fiber optic cables, ensuring compatibility with the intended application and transmission distance.

Moreover, fiber optic cables are available in various fiber counts, ranging from a single fiber to hundreds or thousands of fibers bundled together in a single cable assembly. Fiber count is a critical factor in determining the capacity and scalability of fiber optic networks, with higher fiber counts enabling greater bandwidth and flexibility for future expansion. CLI commands such as "fiber-optic-cable-fiber-count-check" are used to check the fiber count of fiber optic cables, ensuring they meet the capacity requirements of the network infrastructure.

Furthermore, fiber optic cables are rated for different environmental conditions, including indoor, outdoor, and underground installations. Indoor cables are typically designed for use within buildings or controlled environments, with jacket materials optimized for flame resistance, low smoke emission, and compatibility with building codes and regulations. Outdoor cables are designed for installation in outdoor environments, with jacket materials and construction techniques optimized for durability, weather resistance, and UV protection. Underground cables are designed for burial in underground conduits or trenches, with jacket materials and armor layers providing additional protection against moisture, soil corrosion, and mechanical stress. CLI commands such as "fiber-optic-cable-environmental-rating-check" are used to check the environmental rating of fiber optic cables, ensuring they are suitable for the intended installation location and conditions.

In summary, the construction and characteristics of fiber optic cables play a crucial role in determining their performance, reliability, and suitability for different applications and deployment environments. By understanding the composition, mode of propagation, fiber count, and environmental rating of fiber optic cables, network engineers and installers can select the appropriate cables and deploy them effectively to meet the communication needs of modern telecommunications networks.

Chapter 4: Components of a Fiber Optic System

Detectors and receivers are essential components in fiber optic communication systems, responsible for converting optical signals into electrical signals for processing and data transmission. A variety of detector technologies are used in fiber optic receivers, each offering unique characteristics in terms of sensitivity, bandwidth, speed, and noise performance. One of the most common types of detectors used in fiber optic receivers is the photodiode, which operates based on the principle of the photoelectric effect. Photodiodes convert incident photons into electron-hole pairs, generating a photocurrent proportional to the intensity of the incident light. CLI commands such as "photodiode-receiver-install" are used to install and configure photodiode-based receivers in fiber optic communication systems, ensuring proper alignment and sensitivity for signal detection.

In addition to photodiodes, avalanche photodiodes (APDs) are another type of detector commonly used in fiber optic receivers, particularly in long-haul and high-speed communication systems. APDs utilize avalanche multiplication to achieve higher sensitivity and lower noise compared to conventional photodiodes, making them ideal for detecting weak optical signals over long distances. CLI commands such as "APD-receiver-configure" are used to configure and optimize avalanche photodiode receivers for maximum sensitivity and signal-to-noise ratio in fiber optic communication systems.

Moreover, coherent receivers are advanced devices used in coherent optical communication systems to detect and demodulate phase and amplitude-modulated optical signals with high accuracy and efficiency. Coherent receivers typically employ optical local oscillators and digital signal processing (DSP) algorithms to recover the transmitted data from the received optical signal. CLI commands such as "coherent-receiver-tune" are used to tune and optimize coherent receivers for specific modulation formats and data rates in coherent optical communication systems, ensuring accurate and reliable data detection.

Furthermore, PIN photodiodes are another type of detector used in fiber optic receivers, offering a balance between sensitivity, speed, and cost-effectiveness. PIN photodiodes have a larger active area compared to conventional photodiodes, allowing them to capture more incident photons and generate higher photocurrents. PIN photodiodes are commonly used in short-haul and medium-haul communication systems, such as fiber-to-the-home (FTTH) networks and local area networks (LANs), where cost and performance are primary considerations. CLI commands such as "PIN-photodiode-receiver-deploy" are used to deploy and configure PIN photodiode receivers in fiber optic communication systems, ensuring optimal performance and reliability.

Additionally, coherent detection techniques, such as homodyne detection and heterodyne detection, are used in coherent optical receivers to achieve high sensitivity and immunity to noise and distortion. Homodyne detection involves mixing the received optical signal with a local oscillator signal at the same frequency, while heterodyne detection involves mixing the received signal with a local oscillator signal at a slightly different frequency. CLI commands such as "coherent-receiver-detection-mode-select" are used to select and configure the detection mode in coherent optical receivers, ensuring optimal performance and signal recovery in coherent optical communication systems.

Moreover, direct detection receivers are used in intensity-modulated optical communication systems to detect changes in the intensity of the received optical signal directly. Direct detection receivers typically consist of a photodiode followed by electronic amplifiers and signal processing circuits to recover the transmitted data from the received optical signal. CLI commands such as "direct-detection-receiver-configure" are used to configure and optimize direct detection receivers for specific modulation formats and data rates in intensity-modulated optical communication systems, ensuring accurate and reliable data detection.

Furthermore, coherent optical receivers are used in coherent optical communication systems to detect and demodulate phase and amplitude-modulated optical signals with high accuracy and efficiency. Coherent receivers typically employ optical local oscillators and digital signal processing (DSP) algorithms to recover the transmitted data from the received optical signal. CLI commands such as "coherent-optical-receiver-configure" are used to configure and optimize coherent optical receivers for specific modulation formats and data rates in coherent optical communication systems, ensuring accurate and reliable data detection.

In summary, detectors and receivers play a critical role in fiber optic communication systems, converting optical signals into electrical signals for processing and data transmission. From photodiodes and avalanche photodiodes to PIN photodiodes and coherent receivers, each type of detector offers unique characteristics and capabilities for different applications and deployment scenarios. By understanding the properties and performance characteristics of various detectors and receivers, engineers and network operators can design and deploy fiber optic communication systems that meet the demands of modern telecommunications networks.

Detectors and receivers stand as indispensable elements within the realm of fiber optic communication systems, tasked with the conversion of optical signals into electrical ones, facilitating signal processing and data transmission. A myriad of detector technologies, each with its unique characteristics in terms of sensitivity, bandwidth, speed, and noise performance, are employed in fiber optic receivers. Among the prevalent detector types is the photodiode, operating on the principle of the photoelectric effect, converting incident photons into electron-hole pairs, thereby generating a photocurrent proportional to the incident light's intensity. CLI commands such as "install photodiode-receiver" are integral for configuring and setting up photodiode-based receivers, ensuring accurate alignment and sensitivity for optimal signal detection.

In tandem with photodiodes, avalanche photodiodes (APDs) represent another widely utilized detector type, particularly in long-haul and high-speed communication systems. APDs leverage avalanche multiplication to attain heightened sensitivity and diminished noise levels compared to conventional photodiodes, rendering them ideal for detecting faint optical signals over extended distances. CLI commands such as "configure APD-receiver" are essential for fine-tuning and optimizing avalanche photodiode receivers, maximizing sensitivity and signal-to-noise ratio in fiber optic communication systems.

Moreover, coherent receivers, employed in coherent optical communication systems, are tasked with the detection and demodulation of phase and amplitude-modulated optical signals with precision and efficiency. These receivers typically incorporate optical local oscillators and digital signal processing (DSP) algorithms to recover transmitted data from the received optical signal. CLI commands such as "tune coherent-receiver" are pivotal for configuring and optimizing coherent receivers to match specific modulation formats and data rates in coherent optical communication systems, thereby ensuring accurate and reliable data detection.

Furthermore, PIN photodiodes represent another critical detector type in fiber optic receivers, offering a balance between sensitivity, speed, and cost-effectiveness. PIN photodiodes, featuring a larger active area compared to conventional photodiodes, capture more incident photons, generating higher photocurrents. These photodiodes find common application in short-haul and medium-haul communication systems, such as fiber-to-the-home (FTTH) networks and local area networks (LANs), where cost and performance are primary concerns. CLI commands such as "deploy PIN-photodiode-receiver" play a vital role in the deployment and configuration of PIN photodiode receivers, ensuring optimal performance and reliability.

Additionally, coherent detection techniques, including homodyne detection and heterodyne detection, are employed in coherent optical receivers to achieve high sensitivity and immunity to noise and distortion. Homodyne detection involves the mixing of the received optical signal with a local oscillator signal at the same frequency, while heterodyne detection involves mixing the received signal with a local oscillator signal at a slightly different frequency. CLI commands such as "select detection mode-coherent-receiver" are indispensable for choosing and configuring the detection mode in coherent optical receivers, ensuring optimal performance and signal recovery.

Furthermore, direct detection receivers, prevalent in intensity-modulated optical communication systems, detect changes in the intensity of the received optical signal directly. These receivers typically comprise a photodiode followed by electronic amplifiers and signal processing circuits to extract transmitted data from the received optical signal. CLI commands such as "configure direct-detection-receiver" are vital for configuring and optimizing direct detection receivers for specific modulation formats and data rates in intensity-modulated optical communication systems, ensuring accurate and reliable data detection.

Additionally, coherent optical receivers, employed in coherent optical communication systems, detect and demodulate phase and amplitude-modulated optical signals with precision and efficiency. These receivers typically feature optical local oscillators and digital signal processing (DSP) algorithms to recover transmitted data from the received optical signal. CLI commands such as "configure coherent-optical-receiver" play a crucial role in configuring and optimizing coherent optical receivers for specific modulation formats and data rates in coherent optical communication systems, ensuring accurate and reliable data detection.

In summary, detectors and receivers play pivotal roles in fiber optic communication systems, facilitating the conversion of optical signals into electrical ones for signal processing and data transmission. From photodiodes and avalanche photodiodes to PIN photodiodes and coherent receivers, each detector type offers unique characteristics and capabilities tailored for various applications and deployment scenarios. By comprehending the properties and performance attributes of diverse detectors and receivers, engineers and network operators can design and implement fiber optic communication systems adept at meeting the demands of modern telecommunications networks.

Chapter 5: Transmission Modes: Single-mode vs. Multi-mode

Single-mode fiber (SMF), also known as monomode fiber, is a type of optical fiber designed to carry a single mode of light propagation. It is characterized by a small core diameter, typically around 9 micrometers, which allows only one mode of light to propagate through the fiber. This property enables SMF to provide low dispersion and attenuation, making it ideal for long-distance transmission of optical signals with minimal signal loss. CLI commands such as "single-mode-fiber-characterize" are used to characterize the optical properties of single-mode fiber, including core diameter, numerical aperture, and attenuation coefficient, ensuring compatibility with optical communication systems.

One of the key characteristics of single-mode fiber is its low dispersion, which refers to the spreading of optical signals as they propagate through the fiber. SMF exhibits lower dispersion compared to multi-mode fiber (MMF), allowing for higher data rates and longer transmission distances without the need for signal regeneration. This makes SMF suitable for high-speed data transmission applications, such as long-haul telecommunications networks, submarine cables, and intercontinental fiber optic links. CLI commands such as "single-mode-fiber-dispersion-measure" are used to measure and analyze the dispersion characteristics of single-mode fiber, ensuring optimal performance in high-speed transmission systems.

Moreover, single-mode fiber offers low attenuation, which refers to the loss of optical power as light travels through the fiber. SMF has lower attenuation compared to MMF, enabling signals to travel over longer distances without significant loss of signal strength. This property makes SMF well-suited for long-haul transmission applications where minimizing signal loss is critical for maintaining signal integrity and ensuring reliable data transmission. CLI commands such as "single-mode-fiber-attenuation-test" are used to test and verify the attenuation characteristics of single-mode fiber, ensuring that it meets the requirements for long-distance transmission systems.

Furthermore, single-mode fiber supports higher bandwidth compared to multi-mode fiber, allowing for greater data transmission capacity and more channels to be multiplexed onto a single fiber. This increased bandwidth enables higher data rates and greater network scalability, making SMF an ideal choice for high-capacity communication networks, such as backbone networks, metropolitan area networks (MANs), and data center interconnects. CLI commands such as "single-mode-fiber-bandwidth-assess" are used to assess the bandwidth capacity of single-mode fiber and determine its suitability for high-speed data transmission applications.

Additionally, single-mode fiber is immune to modal dispersion, a phenomenon that occurs in multi-mode fiber due to the different propagation paths of light rays within the fiber core. Modal dispersion limits the achievable data rates and transmission distances in MMF systems, particularly at higher data rates and longer wavelengths. By supporting only a single mode of light propagation, SMF eliminates modal dispersion, allowing for higher data rates and longer transmission distances without the need for mode conditioning or dispersion compensation techniques. CLI commands such as "single-mode-fiber-immunity-test" are used to test the immunity of single-mode fiber to modal dispersion, ensuring reliable performance in high-speed transmission systems.

Moreover, single-mode fiber is well-suited for wavelength-division multiplexing (WDM) applications, where multiple optical signals are transmitted simultaneously over different wavelengths within the same fiber. SMF allows for precise control and manipulation of individual wavelengths, enabling dense packing of channels and maximizing the utilization of the available optical spectrum. This enables higher data rates and greater network capacity compared to multi-mode fiber, making SMF the preferred choice for WDM systems deployed in long-haul and high-capacity communication networks. CLI commands such as "single-mode-fiber-WDM-deploy" are used to deploy and configure single-mode fiber for wavelength-division multiplexing applications, ensuring efficient utilization of the optical spectrum and maximizing network capacity.

Furthermore, single-mode fiber is widely used in fiber-to-the-home (FTTH) and fiber-to-the-premises (FTTP) deployments, where high-speed broadband services are delivered to residential and commercial customers. SMF enables the transmission of high-definition video, voice, and data signals over long distances, providing reliable and high-performance connectivity to end-users. CLI commands such as "single-mode-fiber-FTTH-install" are used to install and configure single-mode fiber networks for residential and commercial broadband services, ensuring reliable and high-speed connectivity for end-users.

In summary, single-mode fiber offers numerous advantages, including low dispersion, low attenuation, high bandwidth, immunity to modal dispersion, and compatibility with wavelength-division multiplexing. These characteristics make SMF well-suited for a wide range of applications, including long-haul telecommunications, high-speed data transmission, wavelength-division multiplexing, and fiber-to-the-home deployments. By understanding the characteristics and applications of single-mode fiber, network operators and engineers can design and deploy robust and high-performance fiber optic communication networks to meet the growing demands of modern telecommunications infrastructure.

Multi-mode fiber (MMF) is a type of optical fiber that allows multiple modes of light to propagate through the fiber core simultaneously. It is characterized by a larger core diameter, typically around 50 or 62.5 micrometers, which enables multiple light rays to travel through the fiber at different angles. CLI commands such as "multi-mode-fiber-characterize" can be used to characterize the optical properties of multi-mode fiber, including core diameter, numerical aperture, and modal bandwidth, ensuring compatibility with optical communication systems.

One of the key characteristics of multi-mode fiber is its high modal bandwidth, which refers to the ability of the fiber to support multiple transmission modes simultaneously. MMF can accommodate a larger number of modes compared to single-mode fiber, allowing for higher data rates and greater bandwidth capacity. This makes MMF well-suited for short-distance communication applications, such as local area networks (LANs), campus networks, and fiber-to-the-desk (FTTD) installations. CLI commands such as "multi-mode-fiber-bandwidth-test" can be used to test and verify the modal bandwidth of multi-mode fiber, ensuring optimal performance in high-speed transmission systems.

Moreover, multi-mode fiber offers lower cost and easier installation compared to single-mode fiber, making it a cost-effective choice for premises cabling and short-haul communication links. MMF is less sensitive to alignment errors and imperfections in connectors and splices, allowing for simpler and less expensive installation procedures. This makes MMF ideal for use in enterprise networks, data centers, and building backbones, where cost and ease of deployment are primary considerations. CLI commands such as "multi-mode-fiber-install" are used to install and configure multi-mode fiber networks, ensuring reliable and high-performance connectivity for end-users.

Furthermore, multi-mode fiber supports a wider range of light sources and wavelengths compared to single-mode fiber, allowing for greater flexibility in network design and deployment. MMF can accommodate light sources with larger spectral widths and broader wavelength ranges, including light-emitting diodes (LEDs) and vertical-cavity surface-emitting lasers (VCSELs). This enables the use of cost-effective light sources and simplifies the integration of optical components in multi-mode fiber networks. CLI commands such as "multi-mode-fiber-light-source-select" can be used to select and configure compatible light sources for use with multi-mode fiber, ensuring optimal performance and compatibility in optical communication systems.

Additionally, multi-mode fiber is well-suited for applications that require high-speed data transmission over short distances, such as data center interconnects, storage area networks (SANs), and high-performance computing (HPC) clusters. MMF enables the transmission of high-bandwidth data over short distances without the need for costly signal regeneration or dispersion compensation techniques. This makes MMF an ideal choice for high-density, high-throughput applications where minimizing latency and maximizing throughput are critical. CLI commands such as "multi-mode-fiber-data-center-deploy" can be used to deploy and configure multi-mode fiber networks in data center environments, ensuring efficient and reliable connectivity for networked storage and computing resources.

Moreover, multi-mode fiber is widely used in fiber-to-the-x (FTTx) deployments, where high-speed broadband services are delivered to residential and commercial customers. MMF enables the transmission of high-definition video, voice, and data signals over short distances, providing reliable and high-performance connectivity to end-users. CLI commands such as "multi-mode-fiber-FTTx-install" are used to install and configure multi-mode fiber networks for residential and commercial broadband services, ensuring reliable and high-speed connectivity for end-users.

In summary, multi-mode fiber offers numerous advantages, including high modal bandwidth, lower cost, easier installation, compatibility with a wide range of light sources, and suitability for short-distance communication applications. These characteristics make MMF well-suited for a variety of applications, including LANs, FTTx deployments, data center interconnects, and enterprise networks. By understanding the characteristics and applications of multi-mode fiber, network operators and engineers can design and deploy robust and cost-effective fiber optic communication networks to meet the growing demands of modern telecommunications infrastructure.

Chapter 6: Fiber Optic Connectors and Splicing Techniques

Fiber optic connectors play a crucial role in the field of optical communication, facilitating the connection and termination of optical fibers to transmit data reliably and efficiently. There are various types of fiber optic connectors available, each designed for specific applications, environments, and performance requirements. One of the most commonly used types of fiber optic connectors is the SC (Subscriber Connector) connector, which features a push-pull mechanism for easy and secure mating. The SC connector is widely used in telecommunications networks, data centers, and cable television (CATV) systems due to its simplicity, reliability, and compatibility with single-mode and multi-mode fibers. CLI commands such as "install SC-fiber-optic-connector" are used to install and terminate SC connectors on optical fibers, ensuring proper alignment and connectivity for data transmission.

Another popular type of fiber optic connector is the LC (Lucent Connector) connector, which features a small form factor and a snap-in mechanism for quick and easy mating. The LC connector is commonly used in high-density applications, such as data center interconnects, where space-saving and efficient cable management are essential. CLI commands such as "install LC-fiber-optic-connector" are used to install and terminate LC connectors on optical fibers, ensuring precise alignment and reliable connectivity for high-speed data transmission.

Moreover, the ST (Straight Tip) connector is a bayonet-style connector that features a twist-lock mechanism for secure mating. The ST connector is commonly used in local area networks (LANs), fiber-to-the-desk (FTTD) installations, and industrial applications due to its robustness and durability. CLI commands such as "install ST-fiber-optic-connector" are used to install and terminate ST connectors on optical fibers, ensuring reliable and stable connections in harsh environments.

Furthermore, the MTP/MPO (Multi-fiber Push-on/Pull-off) connector is a multi-fiber connector that allows for the simultaneous connection of multiple fibers in a single connector. The MTP/MPO connector is commonly used in high-density applications, such as data center backbone links, where maximizing port density and minimizing installation time are critical. CLI commands such as "install MTP-fiber-optic-connector" are used to install and terminate MTP/MPO connectors on optical fibers, ensuring efficient and reliable connectivity for high-capacity data transmission.

Additionally, the FC (Fiber Connector) connector is a screw-type connector that features a threaded coupling mechanism for secure mating. The FC connector is commonly used in laboratory and test equipment, as well as in military and aerospace applications, due to its ruggedness and stability. CLI commands such as "install FC-fiber-optic-connector" are used to install and terminate FC connectors on optical fibers, ensuring reliable and high-performance connections in demanding environments.

Moreover, the DIN (Deutsches Institut für Normung) connector is a round connector that features a push-pull mechanism for quick and easy mating. The DIN connector is commonly used in industrial and medical applications, as well as in automotive and aerospace systems, due to its compact size and high reliability. CLI commands such as "install DIN-fiber-optic-connector" are used to install and terminate DIN connectors on optical fibers, ensuring dependable connectivity in challenging environments.

Furthermore, the MT-RJ (Mechanical Transfer Registered Jack) connector is a duplex connector that features a small form factor and a push-pull mechanism for easy mating. The MT-RJ connector is commonly used in local area networks (LANs), fiber-to-the-desk (FTTD) installations, and residential applications due to its compact size and simplicity. CLI commands such as "install MT-RJ-fiber-optic-connector" are used to install and terminate MT-RJ connectors on optical fibers, ensuring efficient and reliable connectivity for data transmission.

In summary, there are various types of fiber optic connectors available, each offering unique characteristics and advantages for specific applications and environments. From the SC, LC, ST, and MTP/MPO connectors to the FC, DIN, and MT-RJ connectors, the choice of connector depends on factors such as space constraints, installation requirements, performance specifications, and cost considerations. By understanding the characteristics and applications of different types of fiber optic connectors, network operators and engineers can select the most suitable connectors for their specific needs and deploy them effectively to ensure reliable and efficient optical communication.

Splicing methods and equipment are essential components of fiber optic communication systems, enabling the permanent joining of optical fibers to create continuous and seamless connections for data transmission. There are several splicing methods used in fiber optic networks, each with its unique characteristics, advantages, and applications. One of the most commonly used splicing methods is fusion splicing, which involves permanently fusing the ends of two optical fibers together using localized heat generated by an electric arc. Fusion splicing offers low insertion loss, high tensile strength, and excellent reliability, making it ideal for long-haul transmission applications where signal integrity is critical. CLI commands such as "fusion-splicing-setup" are used to set up fusion splicing equipment, including fusion splicers, cleavers, and fiber holders, ensuring precise alignment and optimal fusion parameters for reliable splicing.

Another widely used splicing method is mechanical splicing, which involves aligning and securing the ends of two optical fibers using a precision-machined splice sleeve or alignment fixture. Mechanical splicing does not involve the fusion of fibers and relies on physical contact between the fiber ends to maintain optical continuity. Mechanical splicing offers quick and easy installation, low cost, and reusability, making it suitable for temporary or emergency repairs, as well as for field installations where fusion splicing equipment may not be available. CLI commands such as "mechanical-splicing-setup" are used to set up mechanical splicing equipment, including splice sleeves, alignment fixtures, and fiber holders, ensuring accurate alignment and reliable splicing.

Moreover, splice-on connectors (SOCs) are a type of fusion splice that combines the benefits of fusion splicing and connectorization, allowing for the direct attachment of pre-polished connectors to optical fibers using fusion splicing equipment. SOCs eliminate the need for field termination and polishing of connectors, reducing installation time and labor costs. SOCs are commonly used in high-density applications, such as data center interconnects, where space-saving and efficient cable management are essential. CLI commands such as "SOC-attachment" are used to attach splice-on connectors to optical fibers using fusion splicing equipment, ensuring precise alignment and reliable connectivity for data transmission.

Furthermore, mass fusion splicing is a specialized splicing technique that allows for the simultaneous fusion splicing of multiple optical fibers using a mass fusion splicer. Mass fusion splicing is commonly used in high-density backbone networks, such as long-haul telecommunications networks and submarine cables, where large numbers of fibers need to be spliced quickly and efficiently. Mass fusion splicing equipment can splice up to 144 fibers simultaneously, significantly reducing installation time and labor costs compared to traditional fusion splicing methods. CLI commands such as "mass-fusion-splicing-setup" are used to set up mass fusion splicing equipment, ensuring proper alignment and fusion parameters for reliable splicing of multiple fibers.

Additionally, ribbon splicing is a specialized splicing technique used for splicing ribbon cables, which consist of multiple fibers arranged in a flat ribbon configuration. Ribbon splicing equipment allows for the simultaneous fusion splicing of all fibers in a ribbon, resulting in quick and efficient splicing with minimal labor and material costs. Ribbon splicing is commonly used in high-density applications, such as fiber-to-the-home (FTTH) deployments and fiber distribution hubs, where space-saving and efficient cable management are essential. CLI commands such as "ribbon-splicing-setup" are used to set up ribbon splicing equipment, ensuring proper alignment and fusion parameters for reliable splicing of ribbon cables.

Moreover, arc fusion splicing is a specialized splicing technique used for splicing specialty fibers, such as polarization-maintaining fibers (PMFs) and erbium-doped fibers (EDFs), which require precise alignment and fusion parameters for optimal performance. Arc fusion splicing equipment offers enhanced control over fusion parameters, such as arc power, duration, and alignment, allowing for the splicing of specialty fibers with high accuracy and repeatability. Arc fusion splicing is commonly used in applications where precise control over splicing parameters is critical, such as fiber laser manufacturing, optical sensing, and biomedical imaging. CLI commands such as "arc-fusion-splicing-setup" are used to set up arc fusion splicing equipment, ensuring precise alignment and fusion parameters for reliable splicing of specialty fibers.

In summary, splicing methods and equipment play a crucial role in fiber optic communication systems, enabling the creation of continuous and seamless connections for data transmission. From fusion splicing and mechanical splicing to splice-on connectors, mass fusion splicing, ribbon splicing, and arc fusion splicing, each splicing technique offers unique advantages and applications for different deployment scenarios. By understanding the characteristics and capabilities of various splicing methods and equipment, network operators and engineers can select the most suitable splicing technique for their specific needs and deploy it effectively to ensure reliable and efficient optical communication.

Chapter 7: Signal Loss and Dispersion in Fiber Optics

Attenuation and absorption are fundamental concepts in the field of fiber optic communication, governing the loss of optical power as light propagates through an optical fiber. Attenuation refers to the reduction in the intensity of an optical signal as it travels along the fiber, primarily due to factors such as scattering, bending losses, and material impurities. CLI commands such as "measure-attenuation" are used to measure the attenuation of optical fibers, providing valuable insights into the performance and reliability of fiber optic communication systems.

One of the primary mechanisms contributing to attenuation in optical fibers is scattering, which occurs when light interacts with microscopic imperfections in the fiber core, such as variations in refractive index or density. CLI commands such as "analyze-scattering-effects" are used to analyze the scattering effects in optical fibers, identifying potential sources of attenuation and optimizing fiber design and manufacturing processes to minimize scattering losses.

Moreover, bending losses occur when an optical fiber is bent beyond its critical bending radius, causing light to leak out of the fiber core and into the cladding or surrounding medium. CLI commands such as "calculate-bending-loss" are used to calculate the bending loss of optical fibers, taking into account factors such as fiber diameter, curvature, and material properties to ensure proper fiber handling and installation practices.

Furthermore, material impurities such as water vapor, hydrogen, and metallic ions can contribute to attenuation by absorbing and scattering light as it propagates through the fiber. CLI commands such as "analyze-material-impurities" are used to analyze the effects of material impurities on optical fiber performance, identifying sources of contamination and implementing measures to minimize their impact on attenuation.

In addition to attenuation, absorption is another important mechanism that contributes to the loss of optical power in fiber optic communication systems. Absorption occurs when light is absorbed by the atoms and molecules in the fiber core and cladding, converting optical energy into heat or other forms of energy. CLI commands such as "measure-absorption-coefficient" are used to measure the absorption coefficient of optical fibers, quantifying the rate at which light is absorbed as it travels through the fiber.

One of the main contributors to absorption in optical fibers is intrinsic absorption, which arises from the absorption bands of materials used in fiber fabrication, such as silica glass. CLI commands such as "analyze-intrinsic-absorption" are used to analyze the intrinsic absorption properties of optical fibers, identifying absorption peaks and optimizing fiber composition and manufacturing processes to minimize intrinsic absorption losses.

Moreover, extrinsic absorption can occur due to impurities and dopants introduced during the fiber fabrication process, such as transition metal ions or rare earth dopants. CLI commands such as "analyze-extrinsic-absorption" are used to analyze the extrinsic absorption effects in optical fibers, identifying sources of contamination and implementing purification and doping control measures to minimize extrinsic absorption losses.

Furthermore, absorption losses can also arise from nonlinear effects such as two-photon absorption and free carrier absorption, which become significant at high optical power levels or in special fiber designs. CLI commands such as "simulate-nonlinear-absorption" are used to simulate the nonlinear absorption effects in optical fibers, predicting the impact of nonlinear phenomena on system performance and implementing mitigation strategies such as power control and dispersion management.

Additionally, absorption losses can be minimized through proper fiber design and material selection, as well as by employing techniques such as fiber doping, core-cladding index matching, and fiber coating optimization. CLI commands such as "optimize-fiber-design" are used to optimize the design of optical fibers, ensuring low absorption and attenuation for maximum signal transmission efficiency.

In summary, attenuation and absorption are critical factors that affect the performance and reliability of fiber optic communication systems. By understanding the mechanisms and sources of attenuation and absorption in optical fibers, network operators and engineers can implement measures to minimize losses and optimize system performance for reliable and efficient data transmission. CLI commands play a crucial role in analyzing, measuring, and optimizing attenuation and absorption properties, providing valuable insights into fiber optic communication system design and deployment.

Chromatic and modal dispersion are two important phenomena that can affect the performance of fiber optic communication systems, particularly in high-speed data transmission applications. Chromatic dispersion refers to the spreading of optical signals due to the variation in the speed of light with respect to different wavelengths, while modal dispersion arises from the different propagation paths of light rays within a multi-mode optical fiber. CLI commands such as "measure-chromatic-dispersion" and "measure-modal-dispersion" are used to measure the chromatic and modal dispersion of optical fibers, respectively, providing valuable insights into system performance and reliability.

Chromatic dispersion occurs because the refractive index of an optical fiber varies slightly with wavelength, causing different wavelengths of light to travel at slightly different speeds. This leads to the dispersion of optical signals over long distances, limiting the achievable data rates and transmission distances in fiber optic communication systems. CLI commands such as "analyze-chromatic-dispersion-profile" are used to analyze the chromatic dispersion profile of optical fibers, identifying dispersion peaks and optimizing fiber design and transmission parameters to minimize chromatic dispersion effects.

Moreover, chromatic dispersion can be compensated for using techniques such as dispersion-shifted fibers, dispersion compensation modules, and fiber Bragg gratings. Dispersion-shifted fibers are designed to minimize chromatic dispersion by shifting the zero-dispersion wavelength to the desired operating wavelength, while dispersion compensation modules and fiber Bragg gratings are used to selectively compensate for chromatic dispersion at specific wavelengths. CLI commands such as "deploy-dispersion-compensation-module" are used to deploy dispersion compensation modules in fiber optic communication systems, ensuring optimal performance and reliability.

Furthermore, modal dispersion occurs in multi-mode optical fibers due to the different propagation paths of light rays, resulting in the spreading of optical signals over long distances. Modal dispersion limits the achievable data rates and transmission distances in multi-mode fiber optic communication systems, particularly at higher data rates and longer wavelengths. CLI commands such as "analyze-modal-dispersion-pattern" are used to analyze the modal dispersion pattern of multi-mode optical fibers, identifying dispersion peaks and optimizing fiber design and transmission parameters to minimize modal dispersion effects.

In addition to optimizing fiber design, modal dispersion can be minimized through the use of graded-index fibers, which have a parabolic refractive index profile that reduces modal dispersion by allowing light rays to travel along curved paths. CLI commands such as "deploy-graded-index-fiber" are used to deploy graded-index fibers in multi-mode fiber optic communication systems, ensuring reduced modal dispersion and improved system performance.

Moreover, modal dispersion can also be compensated for using techniques such as mode conditioning cables and spatial mode multiplexing. Mode conditioning cables are used to condition the modal distribution of light rays entering the fiber, reducing modal dispersion and improving system performance. Spatial mode multiplexing involves multiplexing multiple spatial modes onto a single fiber, effectively reducing modal dispersion and increasing the achievable data rates and transmission distances. CLI commands such as "deploy-mode-conditioning-cable" and "deploy-spatial-mode-multiplexing" are used to deploy mode conditioning cables and spatial mode multiplexing techniques in multi-mode fiber optic communication systems, respectively, ensuring optimal performance and reliability.

Furthermore, advanced signal processing techniques such as digital signal processing (DSP) and coherent detection can be used to mitigate the effects of chromatic and modal dispersion in fiber optic communication systems. DSP algorithms are used to compensate for dispersion-induced distortions in the received signal, while coherent detection techniques enable the recovery of phase information lost due to dispersion effects. CLI commands such as "configure-DSP-algorithm" and "enable-coherent-detection" are used to configure DSP algorithms and enable coherent detection in fiber optic communication systems, ensuring accurate signal recovery and improved system performance.

In summary, chromatic and modal dispersion are important phenomena that can affect the performance and reliability of fiber optic communication systems. By understanding the mechanisms and effects of chromatic and modal dispersion, network operators and engineers can implement measures to minimize dispersion effects and optimize system performance for reliable and efficient data transmission. CLI commands play a crucial role in analyzing, measuring, and mitigating dispersion effects, providing valuable insights into fiber optic communication system design and deployment.

Chapter 8: Fiber Optic Networking: LANs and WANs

Local Area Networks (LANs) are fundamental components of modern computer networks, providing connectivity within a limited geographical area such as a home, office building, or campus. LANs enable the sharing of resources and information among devices such as computers, printers, servers, and storage devices, facilitating collaboration and productivity in various environments. CLI commands such as "configure-LAN-settings" are used to configure LAN settings, including IP addresses, subnet masks, and default gateways, ensuring proper network connectivity and communication.

One of the key components of a LAN is the network interface card (NIC), which enables devices to connect to the network and communicate with other devices. NICs are installed in computers, servers, printers, and other networked devices, providing a physical interface for transmitting and receiving data over the LAN. CLI commands such as "check-NIC-status" are used to check the status of network interface cards, ensuring proper functioning and connectivity.

Moreover, LANs are typically built using Ethernet technology, which defines the standards and protocols for communication between devices on the network. Ethernet uses a combination of hardware and software components, including Ethernet cables, switches, and routers, to transmit data packets between devices. CLI commands such as "configure-Ethernet-switch" are used to configure Ethernet switches, including port settings, VLANs, and trunking, ensuring efficient data transmission and network management.

Furthermore, LANs are often organized into logical segments called subnets, which allow for efficient addressing and routing of data packets within the network. Subnets are defined by their IP address range and subnet mask, which determine the range of addresses available to devices on the subnet. CLI commands such as "create-subnet" are used to create subnets, specifying the IP address range and subnet mask, ensuring proper segmentation and management of network traffic.

In addition to Ethernet technology, LANs may also incorporate wireless networking technologies such as Wi-Fi, which allow devices to connect to the network without the need for physical cables. Wi-Fi networks use radio waves to transmit data between devices, providing flexibility and mobility for users within the coverage area. CLI commands such as "configure-WiFi-network" are used to configure Wi-Fi networks, including SSIDs, security settings, and access controls, ensuring secure and reliable wireless connectivity.

Moreover, LANs often include network infrastructure devices such as switches and routers, which play a crucial role in directing and managing network traffic. Switches are used to connect devices within the same LAN segment, enabling fast and efficient communication between devices. CLI commands such as "configure-network-switch" are used to configure network switches, including port settings, VLANs, and quality of service (QoS) policies, ensuring optimal performance and reliability.

Furthermore, routers are used to connect multiple LANs together and enable communication between devices on different networks. Routers use routing tables and protocols such as IP routing and dynamic routing to determine the best path for data packets to travel between networks. CLI commands such as "configure-network-router" are used to configure network routers, including routing protocols, access control lists (ACLs), and network address translation (NAT), ensuring secure and efficient inter-network communication.

Additionally, LANs may incorporate network services such as DHCP (Dynamic Host Configuration Protocol) and DNS (Domain Name System), which provide essential functions for device configuration and name resolution. DHCP allows devices to automatically obtain IP addresses, subnet masks, and other network parameters from a central server, simplifying network administration and management. CLI commands such as "configure-DHCP-server" are used to configure DHCP servers, specifying address pools, lease durations, and other parameters, ensuring reliable and efficient IP address allocation.

Moreover, DNS translates domain names into IP addresses, allowing devices to locate and communicate with each other using human-readable names instead of numerical IP addresses. DNS servers maintain databases of domain names and their corresponding IP addresses, resolving queries from client devices and directing them to the appropriate destination. CLI commands such as "configure-DNS-server" are used to configure DNS servers, specifying domain zones, records, and forwarders, ensuring accurate and efficient name resolution.

In summary, Local Area Networks (LANs) are essential components of modern computer networks, providing connectivity and communication within a limited geographical area. LANs use Ethernet technology, wireless networking, and network infrastructure devices such as switches and routers to enable the sharing of resources and information among devices. By understanding the basics of LANs and deploying the appropriate technologies and configurations, network administrators can ensure reliable and efficient network connectivity for users and devices.

Wide Area Networks (WANs) play a crucial role in connecting geographically dispersed locations and enabling communication between devices over long distances. Unlike Local Area Networks (LANs), which operate within a limited geographical area, WANs cover vast regions and often span multiple cities, countries, or continents. CLI commands such as "configure-WAN-connection" are used to configure WAN connections, including leased lines, MPLS circuits, and VPN tunnels, ensuring reliable and secure connectivity between remote sites.

One of the key technologies used in WANs is the fiber optic backbone, which provides the high-speed, high-capacity infrastructure necessary to support the transmission of data over long distances. Fiber optic cables use optical fibers made of glass or plastic to transmit data as pulses of light, offering significant advantages over traditional copper cables, including higher bandwidth, lower latency, and immunity to electromagnetic interference. CLI commands such as "deploy-fiber-optic-backbone" are used to deploy fiber optic backbone networks, including laying and splicing fiber optic cables, installing optical amplifiers, and configuring network equipment, ensuring reliable and efficient data transmission.

Moreover, fiber optic backbone networks use a variety of transmission technologies to transmit data over long distances, including wavelength division multiplexing (WDM), dense wavelength division multiplexing (DWDM), and coherent optical communication. WDM technology enables multiple data streams to be transmitted simultaneously over a single fiber optic cable by using different wavelengths of light to carry each data stream. CLI commands such as "configure-WDM-system" are used to configure WDM systems, including wavelength assignments, power levels, and optical filters, ensuring efficient multiplexing and demultiplexing of data streams.

Furthermore, DWDM technology extends the capacity of fiber optic backbone networks by allowing for the transmission of multiple wavelengths of light within the same optical fiber, effectively increasing the data transmission capacity by several orders of magnitude.

CLI commands such as "configure-DWDM-network" are used to configure DWDM networks, including channel spacing, modulation formats, and dispersion compensation, ensuring optimal utilization of optical spectrum and maximizing data throughput.

In addition to WDM and DWDM technologies, coherent optical communication enables the transmission of high-speed data over long distances by using advanced modulation formats and digital signal processing techniques to overcome impairments such as chromatic dispersion and polarization mode dispersion. CLI commands such as "configure-coherent-optical-system" are used to configure coherent optical communication systems, including modulation formats, digital signal processing algorithms, and error correction codes, ensuring reliable and efficient data transmission over fiber optic backbone networks.

Moreover, fiber optic backbone networks often incorporate network infrastructure devices such as optical amplifiers, multiplexers, and demultiplexers, which play a crucial role in amplifying, combining, and separating optical signals as they travel through the network. Optical amplifiers use semiconductor or fiber-based amplification techniques to boost the power of optical signals, enabling data transmission over long distances without the need for frequent regeneration. CLI commands such as "configure-optical-amplifier" are used to configure optical amplifiers, including gain levels, noise figures, and pump powers, ensuring optimal signal amplification and network performance.

Furthermore, multiplexers and demultiplexers are used to combine and separate optical signals at different wavelengths, respectively, enabling the efficient multiplexing and demultiplexing of data streams within the fiber optic backbone network. CLI commands such as "configure-optical-multiplexer" and "configure-optical-demultiplexer" are used to configure multiplexers and demultiplexers, including channel assignments, port configurations, and signal levels, ensuring accurate and efficient signal routing within the network.

Additionally, fiber optic backbone networks may incorporate network management and monitoring systems to ensure the reliable operation and performance of the network. Network management systems provide centralized control and monitoring of network devices, allowing administrators to configure network settings, monitor network performance, and troubleshoot network issues remotely. CLI commands such as "configure-network-management-system" are used to configure network management systems, including user access controls, alarm thresholds, and event logging, ensuring efficient network management and maintenance.

In summary, Wide Area Networks (WANs) and fiber optic backbone networks are essential components of modern telecommunications infrastructure, enabling the transmission of data over long distances with high-speed, high-capacity connectivity. By deploying fiber optic backbone networks and leveraging advanced transmission technologies such as WDM, DWDM, and coherent optical communication, organizations can ensure reliable and efficient communication between geographically dispersed locations. CLI commands play a crucial role in configuring and managing fiber optic backbone networks, providing administrators with the tools they need to deploy, monitor, and maintain WAN connectivity effectively.

Chapter 9: Fiber Optic Testing and Troubleshooting

Testing equipment and procedures are essential components of ensuring the reliability, performance, and quality of fiber optic communication systems. Testing equipment includes a variety of tools and instruments designed to measure various parameters such as optical power, loss, dispersion, and reflectance, while testing procedures involve a series of steps and methodologies for conducting accurate and comprehensive tests on fiber optic networks. CLI commands such as "test-optical-power" and "measure-fiber-loss" are used to perform optical power measurements and fiber loss measurements, respectively, providing valuable insights into the performance and health of fiber optic communication systems.

One of the primary testing procedures in fiber optic networks is optical power testing, which involves measuring the optical power levels of transmitted signals to ensure they meet specified requirements and standards. Optical power meters are used to measure the absolute or relative optical power levels of light signals transmitted through optical fibers, providing critical information about signal strength, attenuation, and losses. CLI commands such as "calibrate-optical-power-meter" are used to calibrate optical power meters, ensuring accurate and reliable measurements.

Moreover, optical time-domain reflectometers (OTDRs) are used to measure fiber optic cable lengths, detect faults, and characterize optical signal reflections and losses along the length of the fiber. OTDRs emit short pulses of light into the fiber and measure the time it takes for the light to be reflected back, allowing for the accurate determination of fiber length and identification of any anomalies or discontinuities in the fiber. CLI commands such as "configure-OTDR-scan" are used to configure OTDR scanning parameters, including pulse width, averaging, and measurement range, ensuring optimal measurement accuracy and resolution. Furthermore, optical spectrum analyzers (OSAs) are used to analyze the spectral characteristics of optical signals, including wavelength, power, and bandwidth, providing valuable insights into signal quality, stability, and integrity. OSAs use diffraction grating or Fourier transform techniques to disperse optical signals into their component wavelengths and measure their corresponding power levels, enabling detailed spectral analysis and troubleshooting of fiber optic networks. CLI commands such as "analyze-optical-spectrum" are used to analyze optical spectra, identifying spectral peaks, noise levels, and modulation formats.

In addition to optical power testing and spectral analysis, fiber optic networks also undergo testing for various types of losses, including insertion loss, return loss, and dispersion-induced loss. Insertion loss refers to the decrease in optical power caused by the insertion of a device or component into the fiber optic link, such as connectors, splices, or attenuators. CLI commands such as "measure-insertion-loss" are used to measure insertion loss, ensuring that devices and components are properly installed and aligned to minimize signal attenuation.

Moreover, return loss, also known as reflectance, refers to the amount of light that is reflected back towards the source due to imperfections or discontinuities in the fiber optic link, such as connectors, splices, or fiber ends. Return loss measurements are critical for assessing signal integrity, identifying sources of reflection, and ensuring proper termination and alignment of fiber optic connectors. CLI commands such as "measure-return-loss" are used to measure return loss, providing insights into the quality and performance of fiber optic connections.

Furthermore, dispersion-induced loss refers to the degradation of signal quality caused by chromatic dispersion, polarization mode dispersion, or modal dispersion in the fiber optic link. Dispersion-induced loss measurements are essential for evaluating the transmission characteristics of optical fibers and optimizing system performance for high-speed data transmission. CLI commands such as "analyze-dispersion-induced-loss" are used to analyze dispersion-induced loss, identifying sources of dispersion and implementing mitigation strategies such as dispersion compensation or fiber replacement.

Additionally, testing equipment and procedures are also used for fiber optic cable certification, which involves verifying that installed fiber optic cables meet specified performance standards and specifications. Cable certification tests include measurements of optical power, attenuation, length, bandwidth, and other parameters to ensure compliance with industry standards such as TIA/EIA-568 and ISO/IEC 11801. CLI commands such as "certify-fiber-optic-cable" are used to certify fiber optic cables, providing documentation and validation of cable performance for warranty and quality assurance purposes.

Moreover, testing equipment such as optical power meters, OTDRs, OSAs, and cable certifiers are commonly used by network installers, technicians, and engineers to verify the performance and reliability of fiber optic communication systems during installation, commissioning, and maintenance activities. Regular testing and monitoring of fiber optic networks are essential for identifying and resolving issues such as signal degradation, cable damage, connector contamination, and equipment malfunction, ensuring optimal network performance and uptime. CLI commands such as "conduct-network-testing" and "analyze-test-results" are used to conduct network testing and analyze test results, respectively, providing valuable insights into network health and performance.

In summary, testing equipment and procedures play a critical role in ensuring the reliability, performance, and quality of fiber optic communication systems. By using a combination of optical power meters, OTDRs, OSAs, and cable certifiers, network operators and technicians can accurately measure and analyze various parameters such as optical power, loss, dispersion, and reflectance, providing insights into network health and performance. CLI commands are essential for configuring testing equipment, conducting tests, and analyzing test results, enabling efficient and effective troubleshooting and maintenance of fiber optic networks.

In the world of fiber optic communication systems, various issues can arise, affecting network performance and reliability. Understanding common issues and having effective troubleshooting techniques in place is essential for maintaining optimal system operation. CLI commands such as "show-interface-status" and "ping" are used to identify interface status and perform connectivity tests, respectively, providing valuable insights into network health and diagnosing potential problems.

One of the most common issues in fiber optic networks is connectivity problems, which can arise due to issues such as faulty cables, damaged connectors, or misconfigured network devices. To troubleshoot connectivity issues, network administrators can use CLI commands such as "show-interface-status" to check the status of network interfaces and "show-interface-config" to verify interface configurations, ensuring proper connectivity settings and identifying any misconfigurations or errors.

Moreover, conducting ping tests using the "ping" command can help diagnose connectivity issues by sending ICMP (Internet Control Message Protocol) echo requests to remote devices and measuring response times. If ping tests fail or show high latency, it may indicate network congestion, packet loss, or routing issues that need to be addressed. CLI commands such as "ping <destination IP>" are used to perform ping tests, providing insights into network reachability and performance.

Furthermore, another common issue in fiber optic networks is signal loss, which can occur due to factors such as attenuation, dispersion, or optical losses in fiber optic cables and components. To troubleshoot signal loss issues, network administrators can use optical power meters to measure the optical power levels of transmitted signals and identify any significant losses or discrepancies. CLI commands such as "measure-optical-power" are used to measure optical power levels, ensuring that signals are transmitted at adequate power levels and diagnosing potential issues affecting signal quality.

In addition to signal loss, another common issue in fiber optic networks is reflectance, which refers to the amount of light that is reflected back towards the source due to imperfections or discontinuities in the fiber optic link. High reflectance levels can cause signal degradation, interference, and signal-to-noise ratio (SNR) issues, affecting network performance and reliability. To troubleshoot reflectance issues, network administrators can use CLI commands such as "measure-return-loss" to measure return loss and identify sources of reflection or impedance mismatches.

Moreover, fiber optic networks may encounter issues related to fiber bending or twisting, which can cause signal distortion, attenuation, and loss. To troubleshoot fiber bending issues, network administrators can visually inspect fiber optic cables for signs of physical damage or stress, such as kinks, bends, or twists, and use CLI commands such as "check-fiber-status" to verify fiber integrity and continuity. If fiber bending issues are identified, cables may need to be repositioned, rerouted, or replaced to restore proper signal transmission.

Furthermore, fiber optic networks may experience issues related to environmental factors such as temperature fluctuations, humidity, and exposure to moisture or contaminants. Environmental issues can affect the performance and reliability of fiber optic cables and components, leading to signal degradation, corrosion, or fiber breakage. To troubleshoot environmental issues, network administrators can use CLI commands such as "monitor-environmental-conditions" to monitor temperature, humidity, and other environmental parameters, ensuring that network infrastructure is properly protected and maintained.

Additionally, fiber optic networks may encounter issues related to equipment malfunction or failure, such as defective transceivers, damaged connectors, or faulty network devices. To troubleshoot equipment issues, network administrators can use CLI commands such as "show-device-status" to check the status of network devices and "debug" commands to enable debugging and diagnostic features on network equipment. By analyzing log files, error messages, and system alerts, administrators can identify and resolve equipment issues efficiently.

Moreover, software-related issues such as configuration errors, firmware bugs, or compatibility issues can also affect the performance and reliability of fiber optic networks. To troubleshoot software issues, network administrators can use CLI commands such as "show-configuration" to view device configurations, "debug" commands to enable debugging features, and "upgrade" commands to install firmware updates or patches. By analyzing configuration settings, log files, and system messages, administrators can diagnose and resolve software-related issues effectively.

Furthermore, network administrators may encounter issues related to security vulnerabilities, such as unauthorized access, data breaches, or malware infections. To troubleshoot security issues, administrators can use CLI commands such as "show-security-status" to view security settings and configurations, "audit" commands to perform security audits and compliance checks, and "monitor" commands to monitor network traffic for suspicious activity. By implementing security best practices, such as access control, encryption, and intrusion detection, administrators can mitigate security risks and protect sensitive data.

In summary, understanding common issues and having effective troubleshooting techniques in place is essential for maintaining the reliability, performance, and security of fiber optic communication systems. By using CLI commands to identify, diagnose, and resolve network issues, administrators can ensure that fiber optic networks operate smoothly and efficiently, providing reliable connectivity and communication for users and devices.

Chapter 10: Future Trends in Fiber Optics

Advances in fiber optic technology have revolutionized telecommunications, enabling faster data transmission, longer transmission distances, and higher bandwidth capacity than ever before. CLI commands such as "show-tech-support" and "configure-fiber-optic-network" are used to gather technical information and deploy new technologies, respectively, ensuring efficient implementation and management of advanced fiber optic systems.

One of the significant advancements in fiber optic technology is the development of coherent optical communication systems, which utilize advanced modulation formats and digital signal processing techniques to overcome signal impairments and maximize data transmission rates. Coherent optical communication systems enable the transmission of high-speed data over long distances with unprecedented reliability and efficiency. CLI commands such as "configure-coherent-optical-system" are used to configure coherent optical communication systems, including modulation formats, error correction codes, and digital signal processing algorithms, ensuring optimal performance and data throughput.

Moreover, another key advancement in fiber optic technology is the deployment of wavelength division multiplexing (WDM) and dense wavelength division multiplexing (DWDM) systems, which allow multiple data streams to be transmitted simultaneously over a single optical fiber using different wavelengths of light. WDM and DWDM systems significantly increase the capacity and efficiency of fiber optic networks, enabling the aggregation of multiple services and applications onto a single fiber infrastructure. CLI commands such as "configure-WDM-system" are used to configure WDM and DWDM systems, specifying wavelength assignments, channel spacing, and optical power levels, ensuring efficient multiplexing and demultiplexing of data streams.

Furthermore, advancements in fiber optic cable design and manufacturing have led to the development of new types of optical fibers with enhanced performance characteristics, such as reduced attenuation, increased bandwidth, and improved signal integrity. Specialty fibers, such as dispersion-shifted fibers, non-zero dispersion-shifted fibers, and photonic crystal fibers, offer unique properties and capabilities that address specific application requirements and challenges. CLI commands such as "install-specialty-optical-fiber" are used to deploy specialty optical fibers, ensuring proper installation and alignment to maximize performance and reliability.

In addition to improved fiber optic cables, advancements in optical amplification technology have led to the development of erbium-doped fiber amplifiers (EDFAs) and Raman amplifiers, which boost optical signal power levels without the need for electrical regeneration. EDFAs and Raman amplifiers enable the extension of transmission distances and the reduction of signal losses in fiber optic networks, allowing for the deployment of long-haul and ultra-long-haul communication links. CLI commands such as "configure-optical-amplifier" are used to configure optical amplifiers, including gain levels, noise figures, and pump powers, ensuring optimal signal amplification and network performance.

Moreover, advances in fiber optic sensing technology have opened up new possibilities for applications in various industries, including oil and gas, aerospace, and healthcare. Fiber optic sensors utilize the principles of light propagation within optical fibers to measure physical parameters such as temperature, pressure, strain, and vibration with high accuracy and sensitivity. Distributed fiber optic sensing systems, such as distributed temperature sensing (DTS) and distributed acoustic sensing (DAS), enable real-time monitoring and detection of events and anomalies over large areas and long distances. CLI commands such as "deploy-fiber-optic-sensor-network" are used to deploy fiber optic sensing systems, including sensor placement, calibration, and data acquisition, ensuring reliable and accurate measurement of physical parameters.

Furthermore, advances in fiber optic network management and control systems have led to the development of intelligent optical networks that can dynamically adapt to changing traffic demands and network conditions. Software-defined networking (SDN) and network function virtualization (NFV) technologies enable the programmable and flexible control of network resources, allowing for the optimization of network performance, efficiency, and reliability. CLI commands such as "configure-SDN-controller" and "deploy-NFV-platform" are used to configure SDN controllers and deploy NFV platforms, respectively, enabling network operators to automate provisioning, orchestration, and management of fiber optic networks. Additionally, advances in fiber optic technology have facilitated the deployment of fiber-to-the-home (FTTH) and fiber-to-the-premises (FTTP) networks, which provide high-speed broadband internet access to residential and commercial users. FTTH and FTTP networks offer significant advantages over traditional copper-based DSL and cable networks, including faster speeds, higher reliability, and greater bandwidth capacity. CLI commands such as "deploy-FTTH-network" and "activate-FTTP-service" are used to deploy and activate fiber optic broadband services, ensuring seamless connectivity and high-quality internet access for end-users.

In summary, advances in fiber optic technology have transformed the telecommunications industry, enabling faster, more reliable, and more efficient communication networks. By leveraging innovations such as coherent optical communication, wavelength division multiplexing, specialty optical fibers, optical amplification, fiber optic sensing, and intelligent network management systems, organizations can build and deploy advanced fiber optic systems that meet the growing demands of modern applications and services. CLI commands play a crucial role in deploying, configuring, and managing these technologies, ensuring optimal performance and reliability of fiber optic networks in diverse environments and applications.

In the rapidly evolving landscape of fiber optic technology, numerous potential applications and innovations continue to emerge, driving advancements across various industries and sectors. These innovations leverage the unique properties of fiber optics, such as high bandwidth, low latency, and immunity to electromagnetic interference, to address diverse challenges and unlock new opportunities for connectivity, communication, and data transmission. CLI commands such as "configure-optical-network" are used to deploy fiber optic networks, specifying parameters such as wavelengths, power levels, and modulation formats to optimize performance and reliability.

One of the most promising applications of fiber optic technology is in telecommunications, where fiber optic networks serve as the backbone for high-speed internet, voice, and video communication services. With the ever-increasing demand for bandwidth-intensive applications such as streaming video, cloud computing, and virtual reality, fiber optic networks offer unparalleled performance and scalability, enabling seamless connectivity and ultra-fast data transmission. CLI commands such as "deploy-fiber-optic-backbone" are used to deploy fiber optic backbone networks, ensuring robust and reliable connectivity for telecommunications providers and their customers.

Moreover, fiber optic technology plays a critical role in enabling the Internet of Things (IoT), where interconnected devices and sensors communicate and exchange data over networked systems. Fiber optic sensors offer advantages such as high sensitivity, immunity to electromagnetic interference, and the ability to operate in harsh environments, making them ideal for monitoring and control applications in industries such as manufacturing, healthcare, and infrastructure. CLI commands such as "install-fiber-optic-sensors" are used to deploy fiber optic sensor networks, connecting sensors to optical fibers and configuring monitoring systems to collect and analyze data in real-time.

Furthermore, fiber optic technology is revolutionizing healthcare and medical imaging, where fiber optic probes and endoscopes enable minimally invasive diagnostic procedures and surgical interventions. Fiber optic probes provide high-resolution imaging and precise localization of tissues and organs, allowing clinicians to visualize internal structures and perform procedures with enhanced precision and accuracy. CLI commands such as "calibrate-fiber-optic-probe" are used to calibrate fiber optic probes, ensuring accurate imaging and diagnosis in medical applications.

Additionally, fiber optic technology is driving innovations in the field of environmental monitoring and sensing, where fiber optic cables and sensors are deployed to monitor parameters such as temperature, pressure, and chemical composition in natural and industrial environments. Fiber optic distributed sensing systems use techniques such as optical time-domain reflectometry (OTDR) and Brillouin optical time-domain analysis (BOTDA) to detect changes in optical signals caused by environmental changes, providing valuable insights into environmental conditions and potential hazards. CLI commands such as "configure-distributed-sensing-system" are used to configure fiber optic distributed sensing systems, specifying parameters such as measurement range, resolution, and sampling rate to optimize performance and accuracy.

Moreover, fiber optic technology is finding applications in smart cities and infrastructure, where fiber optic networks support a wide range of applications such as traffic management, public safety, and utilities monitoring. Fiber optic sensors embedded in roads, bridges, and buildings can detect structural defects, monitor traffic flow, and detect environmental hazards, enabling proactive maintenance and management of urban infrastructure. CLI commands such as "deploy-fiber-optic-sensors" and "configure-smart-city-network" are used to deploy fiber optic sensor networks and smart city infrastructure, ensuring efficient and reliable monitoring and management of urban environments.

Furthermore, fiber optic technology is driving innovations in the field of aerospace and defense, where lightweight and high-performance fiber optic cables and components are used in aircraft, satellites, and military systems. Fiber optic communication systems offer advantages such as low weight, high bandwidth, and immunity to electromagnetic interference, making them ideal for applications such as aircraft avionics, satellite communications, and battlefield networks. CLI commands such as "install-fiber-optic-communication-system" are used to install fiber optic communication systems in aerospace and defense platforms, ensuring secure and reliable communication in challenging environments.

Additionally, fiber optic technology is revolutionizing energy and utilities, where fiber optic networks enable efficient monitoring and control of power grids, pipelines, and renewable energy systems. Fiber optic sensors deployed along power lines and pipelines can detect faults, leaks, and environmental conditions, enabling proactive maintenance and minimizing downtime. CLI commands such as "deploy-fiber-optic-sensor-network" are used to deploy fiber optic sensor networks in energy and utilities infrastructure, ensuring reliable and efficient monitoring and management of critical assets.

In summary, the potential applications and innovations of fiber optic technology are vast and diverse, spanning across industries such as telecommunications, healthcare, environmental monitoring, smart cities, aerospace, defense, and energy. By leveraging the unique properties of fiber optics, such as high bandwidth, low latency, and immunity to electromagnetic interference, organizations can unlock new opportunities for connectivity, communication, and data transmission. CLI commands play a crucial role in deploying fiber optic networks and systems, ensuring optimal performance, reliability, and scalability in various applications and environments.

BOOK 2
MASTERING FIBER OPTIC NETWORKS
ADVANCED TECHNIQUES AND APPLICATIONS

ROB BOTWRIGHT

Chapter 1: Advanced Fiber Optic Cable Designs

High-density cable configurations represent a significant advancement in fiber optic infrastructure design, enabling organizations to maximize capacity and optimize space utilization in data centers, telecommunications networks, and other high-demand environments. CLI commands such as "configure-high-density-cabling" are used to deploy high-density cable configurations, specifying parameters such as cable types, connector types, and installation layouts to achieve efficient and reliable connectivity.

One of the key aspects of high-density cable configurations is the use of compact and space-saving cable designs, such as ribbon cables and micro-cables, which pack a high number of optical fibers into a small footprint. Ribbon cables consist of multiple fiber ribbons stacked side by side, allowing for dense packing and easy termination with ribbon splicing techniques. CLI commands such as "install-ribbon-cables" are used to install ribbon cables, ensuring proper routing and termination for optimal performance.

Moreover, micro-cables are designed with smaller diameters and higher fiber counts than traditional cables, enabling higher packing densities and easier cable management in tight spaces. Micro-cables are often used in applications where space is limited, such as high-density patch panels, fiber distribution frames, and overhead cable trays. CLI commands such as "deploy-micro-cables" are used to deploy micro-cables, ensuring proper installation and routing to minimize signal attenuation and maximize space utilization.

Furthermore, high-density cable configurations often incorporate advanced connector designs, such as MTP/MPO connectors, which allow for the termination of multiple fibers in a single connector, reducing the overall footprint and simplifying cable management. MTP/MPO connectors feature a compact and modular design, with up to 72 fibers terminated in a single connector, enabling rapid deployment and easy scalability in high-density environments. CLI commands such as "terminate-MTP-connectors" are used to terminate MTP/MPO connectors, ensuring precise alignment and reliable connectivity.

In addition to compact cables and connectors, high-density cable configurations utilize innovative cable management techniques to optimize space utilization and minimize cable congestion. Techniques such as cable routing, bundling, and slack management are employed to organize cables and maintain proper bend radii, ensuring optimal signal transmission and minimizing signal loss. CLI commands such as "manage-cable-routing" and "organize-cable-bundling" are used to implement cable management techniques, ensuring neat and efficient cable installations.

Moreover, high-density cable configurations often employ modular patching systems and fiber distribution units (FDUs) to facilitate easy access and reconfiguration of fiber connections. Modular patch panels allow for the rapid deployment and reconfiguration of fiber connections, with interchangeable adapter plates and modular cassettes supporting various connector types and fiber counts. CLI commands such as "configure-patch-panel" are used to configure modular patch panels, specifying adapter plate configurations and fiber assignments for optimal connectivity.

Furthermore, fiber distribution units (FDUs) provide centralized termination and distribution points for fiber optic cables, with pre-terminated trunks and cabling assemblies connecting to individual patch panels or equipment racks. FDUs support high-density cable configurations by consolidating fiber connections and minimizing cable clutter, with options for vertical or horizontal mounting to accommodate different space constraints. CLI commands such as "deploy-FDU" are used to deploy fiber distribution units, ensuring proper alignment and connectivity for efficient fiber distribution.

Additionally, high-density cable configurations often incorporate advanced cable management software and monitoring systems to track cable usage, monitor performance, and identify potential issues in real-time. Cable management software provides centralized control and visibility into cable installations, with features such as asset tracking, port mapping, and cable labeling to simplify maintenance and troubleshooting tasks. CLI commands such as "configure-cable-management-software" are used to configure cable management software, ensuring accurate documentation and efficient management of cable infrastructure.

Moreover, high-density cable configurations enable organizations to future-proof their fiber optic infrastructure and accommodate the growing demand for high-bandwidth applications and services. By deploying compact cables, advanced connectors, innovative cable management techniques, and intelligent monitoring systems, organizations can achieve maximum capacity and efficiency in their fiber optic networks, supporting the needs of today and tomorrow. CLI commands play a crucial role in deploying and managing high-density cable configurations, providing administrators with the tools they need to optimize space utilization, ensure reliable connectivity, and meet the evolving demands of modern communications.

Erbium-Doped Fiber Amplifiers (EDFAs) represent a critical technology in the field of optical communications, providing high-gain amplification for optical signals in fiber optic networks. EDFAs are widely deployed in long-haul transmission systems, metro networks, and submarine cables to compensate for signal losses and extend transmission distances without the need for costly and complex regenerative repeaters. CLI commands such as "configure-EDFA" are used to deploy EDFAs in optical networks, specifying parameters such as gain, output power, and pump power to optimize amplifier performance and ensure reliable signal amplification.

The operation of EDFAs relies on the unique optical properties of erbium-doped fiber, which contains trivalent erbium ions embedded within the core of the optical fiber. When pumped with a laser diode at a specific wavelength, typically around 980 nm or 1480 nm, the erbium ions absorb photons and are raised to higher energy states. CLI commands such as "set-pump-power" are used to set the pump power of the laser diode, ensuring efficient excitation of erbium ions and maximizing amplifier gain.

As the excited erbium ions return to their ground state, they emit photons at wavelengths around 1550 nm, which coincides with the low-loss window of standard single-mode optical fiber. This process, known as stimulated emission, results in the amplification of optical signals propagating through the erbium-doped fiber. CLI commands such as "activate-EDFA" are used to activate EDFAs in optical networks, enabling signal amplification and extending transmission distances.

One of the key advantages of EDFAs is their broad amplification bandwidth, which spans from approximately 1525 nm to 1605 nm, covering the entire C-band of optical communication wavelengths. This wide bandwidth allows EDFAs to amplify multiple wavelength-division multiplexing (WDM) channels simultaneously, enabling dense wavelength-division multiplexing (DWDM) systems with high capacity and spectral efficiency. CLI commands such as "configure-WDM-channels" are used to configure WDM channels in optical networks, specifying wavelengths, power levels, and channel spacing to optimize spectral efficiency and maximize network capacity.

Moreover, EDFAs offer high gain and low noise characteristics, making them ideal for long-haul transmission applications where signal attenuation and noise accumulation are significant challenges. The gain of an EDFA can be adjusted by controlling the pump power and input signal power, allowing for dynamic gain equalization and compensation of optical losses along the transmission path. CLI commands such as "adjust-EDFA-gain" are used to adjust the gain of EDFAs in optical networks, ensuring optimal signal amplification and power balance.

Additionally, EDFAs can be cascaded in series to achieve higher overall gain and extend transmission distances over long-haul fiber optic links. Cascaded EDFAs are commonly used in submarine cables, transoceanic transmission systems, and terrestrial backbone networks to compensate for signal losses and maintain signal integrity over extended distances. CLI commands such as "cascade-EDFAs" are used to cascade multiple EDFAs in optical networks, specifying the number of amplifiers, their placement, and the interconnection topology to achieve the desired signal amplification and transmission performance.

Furthermore, EDFAs can be integrated into optical line terminal (OLT) and optical network terminal (ONT) equipment to provide amplification and signal conditioning functions in passive optical networks (PONs). In PON architectures, EDFAs are deployed at strategic points along the optical distribution network (ODN) to boost signal power and extend reach to subscriber premises. CLI commands such as "integrate-EDFA-PON" are used to integrate EDFAs into PON equipment, ensuring efficient and reliable signal amplification for broadband access applications.

Moreover, EDFAs are essential components in optical add-drop multiplexer (OADM) systems, where they are used to amplify signals at intermediate nodes in WDM networks without disrupting existing traffic. OADMs allow for selective routing and switching of optical signals at wavelength granularity, enabling flexible and efficient network configurations. CLI commands such as "deploy-OADM-EDFA" are used to deploy EDFAs in OADM systems, ensuring seamless integration and reliable signal amplification for dynamic wavelength routing and management.

Additionally, EDFAs play a critical role in optical amplifier-based regenerative systems, where they are used to amplify and regenerate optical signals to overcome transmission impairments and extend transmission distances. Regenerative systems utilize EDFAs in combination with dispersion compensation, optical filtering, and signal regeneration techniques to improve signal quality and mitigate nonlinear effects in long-haul transmission links. CLI commands such as "configure-regenerative-EDFA" are used to configure EDFAs in regenerative systems, specifying parameters such as gain tilt, noise figure, and output power to optimize signal regeneration performance.

In summary, Erbium-Doped Fiber Amplifiers (EDFAs) are indispensable components in modern optical communication networks, providing high-gain amplification for optical signals over extended distances and enabling high-capacity transmission in long-haul, metro, and access networks. By leveraging the unique properties of erbium-doped fiber and advanced pump laser technology, EDFAs offer broad amplification bandwidth, low noise characteristics, and dynamic gain control capabilities, making them versatile and cost-effective solutions for a wide range of applications. CLI commands play a crucial role in deploying, configuring, and managing EDFAs in optical networks, providing administrators with the tools they need to optimize performance, reliability, and scalability in fiber optic infrastructure.

Chapter 2: Optical Amplification and Signal Regeneration

Distributed amplification techniques represent a significant advancement in the field of optical amplification, enabling the amplification of optical signals over long distances with minimal signal degradation. These techniques utilize distributed amplifiers, which consist of multiple amplification stages distributed along the length of the optical fiber, to boost signal power and compensate for signal losses. CLI commands such as "configure-distributed-amplification" are used to deploy distributed amplification techniques, specifying parameters such as gain, bandwidth, and amplification stages to optimize signal performance and quality.

One of the key distributed amplification techniques is distributed Raman amplification, which leverages the Raman effect to amplify optical signals in the fiber itself. In distributed Raman amplification, a high-power pump laser is launched into the fiber at specific wavelengths, stimulating the Raman scattering process and transferring energy from the pump to the signal, resulting in amplification. CLI commands such as "configure-distributed-Raman-amplification" are used to configure distributed Raman amplification systems, specifying pump power levels, wavelengths, and dispersion compensation settings to optimize amplification performance.

Moreover, distributed erbium-doped fiber amplification (EDFA) is another widely used distributed amplification technique, particularly in long-haul telecommunications networks. In distributed EDFA, erbium-doped fiber segments are distributed along the length of the optical fiber, with each segment providing amplification to the optical signal as it propagates through the fiber. CLI commands such as "deploy-distributed-EDFA" are used to deploy distributed EDFA systems, specifying parameters such as pump power, wavelength, and spacing between amplifier segments to achieve optimal signal amplification and performance.

Furthermore, distributed semiconductor optical amplifier (SOA) amplification is gaining prominence as a distributed amplification technique, particularly in high-speed optical communication systems. Distributed SOA amplifiers consist of semiconductor optical amplifiers distributed along the length of the optical fiber, providing amplification to the optical signal through stimulated emission processes. CLI commands such as "configure-distributed-SOA-amplification" are used to configure distributed SOA amplification systems, specifying parameters such as bias current, optical input power, and gain tilt to optimize amplification performance and minimize signal distortion.

Additionally, distributed Brillouin amplification is emerging as a promising distributed amplification technique for enhancing the performance of optical communication systems. In distributed Brillouin amplification, a high-power pump laser generates stimulated Brillouin scattering (SBS) in the fiber, transferring energy from the pump to the signal and providing distributed amplification. CLI commands such as "deploy-distributed-Brillouin-amplification" are used to deploy distributed Brillouin amplification systems, specifying parameters such as pump power, wavelength, and Brillouin gain coefficient to optimize amplification performance and minimize nonlinear effects.

Moreover, hybrid distributed amplification techniques, such as distributed Raman-EDFA amplification and distributed Raman-SOA amplification, combine multiple amplification mechanisms to achieve enhanced performance and flexibility. In distributed Raman-EDFA amplification, Raman amplification provides distributed gain along the fiber length, while EDFA amplifiers are strategically placed to provide additional amplification and compensate for signal losses. CLI commands such as "configure-hybrid-Raman-EDFA-amplification" are used to configure hybrid distributed amplification systems, specifying parameters for both Raman and EDFA amplifiers to optimize amplification performance and achieve desired signal quality.

Furthermore, distributed amplification techniques offer several advantages over traditional lumped amplification techniques, including improved signal-to-noise ratio (SNR), reduced nonlinear effects, and enhanced system flexibility. By distributing amplification along the length of the optical fiber, distributed amplification techniques mitigate signal degradation and distortion, enabling the transmission of high-speed, high-capacity optical signals over long distances without the need for frequent signal regeneration. CLI commands such as "optimize-distributed-amplification" are used to optimize distributed amplification systems, adjusting parameters such as gain profile, pump power distribution, and amplifier spacing to achieve desired performance metrics and meet system requirements.

Additionally, distributed amplification techniques are well-suited for a wide range of applications, including long-haul telecommunications, submarine cable systems, data center interconnects, and high-speed optical networks. By deploying distributed amplification techniques, organizations can enhance the performance and reliability of their optical communication systems, enabling efficient data transmission, increased network capacity, and extended reach. CLI commands play a crucial role in deploying and managing distributed amplification systems, providing administrators with the tools they need to configure, optimize, and monitor amplification performance in real-time.

Distributed amplification techniques represent a pivotal innovation in the realm of optical communication, serving as a cornerstone for long-haul transmission systems and high-capacity networks. These techniques have revolutionized the way optical signals are amplified over extended distances, mitigating signal attenuation and enhancing signal quality through strategic deployment of amplifiers along the fiber span. CLI commands such as "deploy-distributed-amplification" are pivotal in implementing these techniques, enabling network administrators to strategically place amplifiers and optimize signal performance.

One of the most prominent distributed amplification techniques is distributed Raman amplification, which harnesses the Raman effect to amplify optical signals within the fiber itself. CLI commands like "configure-distributed-Raman-amplification" facilitate the configuration of distributed Raman amplification systems, allowing administrators to specify pump power levels, wavelengths, and dispersion compensation settings. Through the stimulation of Raman scattering, energy is transferred from the pump to the signal, resulting in signal amplification and extending the reach of optical transmission systems.

Moreover, distributed erbium-doped fiber amplification (EDFA) stands as another cornerstone in distributed amplification techniques, particularly in telecommunications networks spanning vast distances. CLI commands such as "activate-distributed-EDFA" facilitate the deployment of distributed EDFA systems, enabling administrators to specify parameters like pump power, wavelength, and spacing between amplifier segments. By distributing erbium-doped fiber segments along the fiber span, amplification is achieved incrementally, compensating for signal losses and ensuring signal integrity.

Furthermore, distributed semiconductor optical amplifier (SOA) amplification has gained prominence for its efficacy in high-speed optical communication systems. Distributed SOA amplifiers, deployed strategically along the fiber length, provide amplification through stimulated emission processes. CLI commands such as "install-distributed-SOA-amplification" enable administrators to configure distributed SOA amplification systems, adjusting parameters like bias current, optical input power, and gain tilt to optimize amplification performance and minimize signal distortion.

Additionally, distributed Brillouin amplification emerges as a promising technique for enhancing the performance of optical communication systems. CLI commands such as "deploy-distributed-Brillouin-amplification" empower administrators to configure distributed Brillouin amplification systems, specifying parameters such as pump power, wavelength, and Brillouin gain coefficient. By generating stimulated Brillouin scattering (SBS) within the fiber, energy is transferred from the pump to the signal, resulting in distributed signal amplification and improved system performance.

Moreover, hybrid distributed amplification techniques, such as distributed Raman-EDFA amplification, combine multiple amplification mechanisms to achieve superior performance and flexibility. CLI commands such as "configure-hybrid-Raman-EDFA-amplification" allow administrators to optimize hybrid distributed amplification systems, adjusting parameters for both Raman and EDFA amplifiers. By leveraging the strengths of each amplification technique, hybrid distributed amplification systems offer enhanced performance and reliability in optical communication networks.

Furthermore, distributed amplification techniques offer several advantages over traditional lumped amplification techniques, including improved signal-to-noise ratio (SNR), reduced nonlinear effects, and enhanced system flexibility. By distributing amplification along the length of the optical fiber, distributed amplification techniques mitigate signal degradation and distortion, enabling the transmission of high-speed, high-capacity optical signals over long distances without the need for frequent signal regeneration. CLI commands such as "optimize-distributed-amplification" allow administrators to fine-tune distributed amplification systems, adjusting parameters to achieve desired performance metrics and meet system requirements.

In summary, distributed amplification techniques have revolutionized optical communication systems, enabling efficient transmission of high-capacity data over extended distances. CLI commands play a crucial role in deploying and managing distributed amplification systems, providing administrators with the tools they need to configure, optimize, and monitor amplification performance in real-time. By harnessing the power of distributed amplification techniques, organizations can achieve reliable, high-performance optical communication networks capable of meeting the demands of modern digital connectivity.

Chapter 3: Wavelength Division Multiplexing (WDM)

Wavelength Division Multiplexing (WDM) technology stands as a cornerstone in modern optical communication systems, enabling the transmission of multiple optical signals over a single optical fiber by utilizing different wavelengths or colors of light to carry distinct data streams simultaneously. CLI commands such as "configure-WDM-system" are pivotal in deploying WDM technology, allowing network administrators to set up and manage the multiplexing and demultiplexing of optical signals with precision and efficiency.

At the heart of WDM technology lies the fundamental principle of multiplexing, where multiple optical signals are combined into a single composite signal for transmission over the optical fiber. CLI commands like "create-multiplexed-signal" facilitate the creation of multiplexed signals, allowing administrators to specify the wavelengths or channels to be combined and their corresponding data rates and modulation formats. By multiplexing multiple signals onto a single fiber, WDM technology significantly increases the capacity and efficiency of optical communication systems.

Moreover, WDM technology operates on the principle of wavelength selectivity, where each optical signal is assigned a unique wavelength or color of light that corresponds to its specific data stream. CLI commands such as "assign-wavelengths" enable administrators to assign wavelengths to individual data streams, ensuring that each signal remains distinct and can be efficiently separated at the receiving end. By utilizing different wavelengths, WDM technology allows multiple signals to coexist on the same fiber without interference, maximizing the utilization of the available optical bandwidth.

Furthermore, the principles of WDM technology encompass both coarse wavelength division multiplexing (CWDM) and dense wavelength division multiplexing (DWDM), each offering unique advantages and applications. CLI commands such as "configure-CWDM-system" and "configure-DWDM-system" are used to deploy CWDM and DWDM systems, respectively, specifying parameters such as channel spacing, wavelength range, and optical power levels. CWDM systems utilize wider channel spacing and fewer wavelengths, making them cost-effective solutions for short to medium-haul applications with lower data rates.

In contrast, DWDM systems employ narrower channel spacing and a higher number of wavelengths, enabling the transmission of multiple high-speed data streams over long distances with minimal signal degradation. CLI commands such as "optimize-DWDM-system" are used to optimize DWDM systems, adjusting parameters such as channel spacing, power levels, and dispersion compensation to maximize signal quality and reach. By leveraging the principles of DWDM technology, organizations can achieve high-capacity, long-haul optical communication networks capable of supporting bandwidth-intensive applications.

Moreover, the principles of WDM technology extend to the concept of optical amplification, where optical signals are amplified periodically along the fiber span to compensate for signal losses and maintain signal integrity. CLI commands such as "deploy-optical-amplifiers" facilitate the deployment of optical amplifiers in WDM systems, allowing administrators to specify amplifier types, locations, and power levels. By strategically placing optical amplifiers along the fiber route, WDM systems can achieve extended transmission distances and support large-scale network deployments.

Additionally, the principles of WDM technology encompass wavelength routing and switching, where optical signals are routed and switched based on their wavelengths to their respective destinations. CLI commands such as "configure-wavelength-routing" and "manage-wavelength-switching" enable administrators to configure wavelength routing and switching functions in WDM systems, specifying routing tables, switching matrices, and fault tolerance mechanisms. By dynamically routing and switching wavelengths, WDM systems can optimize network performance, ensure efficient resource utilization, and adapt to changing traffic patterns in real-time.

Furthermore, the principles of WDM technology are underpinned by advanced optical components and subsystems, including optical transceivers, wavelength-selective switches, optical add-drop multiplexers (OADMs), and optical cross-connects (OXCs). CLI commands such as "install-optical-transceivers" and "configure-OADMs" are used to deploy these components in WDM systems, enabling administrators to build flexible and scalable optical networks. By leveraging these advanced components, WDM systems can support diverse applications, from metro and regional networks to long-haul and submarine cable systems.

In summary, the principles of WDM technology revolutionize optical communication systems, enabling high-capacity, high-speed data transmission over long distances with unprecedented efficiency and flexibility. CLI commands play a vital role in deploying and managing WDM systems, providing administrators with the tools they need to configure, optimize, and monitor optical networks with precision and control. By harnessing the principles of WDM technology, organizations can build resilient, future-proof optical communication infrastructure capable of meeting the demands of modern digital connectivity.

Wavelength Division Multiplexing (WDM) technology has revolutionized optical communication networks by allowing multiple optical signals to be transmitted simultaneously over a single optical fiber, each using a different wavelength. CLI commands such as "configure-WDM" are essential in deploying WDM systems, enabling network administrators to allocate wavelengths, configure optical channels, and manage system parameters. There are two primary types of WDM systems: Coarse WDM (CWDM) and Dense WDM (DWDM), each offering distinct advantages and applications in optical networking.

CWDM systems utilize wider channel spacing compared to DWDM systems, typically ranging from 20 to 40 nanometers between adjacent channels. This wider channel spacing simplifies the optical components and reduces the complexity and cost of the system. CLI commands such as "set-CWDM-channel" are used to configure CWDM systems, specifying parameters such as channel frequency and power levels. CWDM systems are well-suited for short to medium-haul applications where cost-effectiveness and simplicity are paramount, such as metropolitan area networks (MANs), access networks, and enterprise deployments.

In contrast, DWDM systems employ much narrower channel spacing, typically on the order of gigahertz or even terahertz, allowing for a significantly higher number of channels to be multiplexed over a single fiber. CLI commands such as "configure-DWDM-channel" are used to configure DWDM systems, specifying parameters such as channel frequency, modulation format, and dispersion compensation. DWDM systems are ideal for long-haul and ultra-long-haul applications where high capacity and scalability are essential, such as backbone networks, submarine cables, and intercontinental links.

One of the key differences between CWDM and DWDM systems lies in their respective channel capacities. CWDM systems typically support up to 18 channels in the 1270 to 1610 nanometer wavelength range, corresponding to the S, C, and L bands of the optical spectrum. CLI commands such as "add-CWDM-channel" are used to add channels to a CWDM system, specifying the wavelength and power levels for each channel. In contrast, DWDM systems can support hundreds or even thousands of channels within the same spectral range, enabling significantly higher aggregate capacities and greater scalability.

Moreover, another differentiating factor between CWDM and DWDM systems is their spectral efficiency. DWDM systems achieve much higher spectral efficiency compared to CWDM systems, allowing for more channels to be multiplexed over the same optical bandwidth. CLI commands such as "optimize-DWDM-spectral-efficiency" are used to optimize DWDM systems for maximum spectral efficiency, adjusting parameters such as channel spacing, modulation format, and dispersion compensation. This higher spectral efficiency enables DWDM systems to achieve higher data rates and accommodate more services and applications within the same optical infrastructure.

Furthermore, DWDM systems offer greater flexibility and granularity in wavelength management compared to CWDM systems. CLI commands such as "tune-DWDM-channel" allow network administrators to dynamically adjust the wavelength and power levels of individual channels in a DWDM system, enabling precise control and optimization of system performance. This flexibility is particularly valuable in dynamic network environments where traffic patterns and capacity requirements may change over time, allowing operators to adapt and optimize the network for changing demands.

Additionally, another key consideration when comparing CWDM and DWDM systems is their reach and dispersion characteristics. CWDM systems typically have higher dispersion tolerance compared to DWDM systems, making them better suited for short to medium-haul applications where dispersion compensation is less critical. CLI commands such as "analyze-CWDM-dispersion" are used to analyze dispersion characteristics in CWDM systems, identifying potential dispersion effects and optimizing system performance. In contrast, DWDM systems may require more sophisticated dispersion compensation techniques, especially in long-haul and ultra-long-haul deployments, to mitigate dispersion effects and ensure signal integrity.

Moreover, another factor to consider is the cost associated with deploying and operating CWDM versus DWDM systems. CWDM systems are generally more cost-effective to deploy and maintain compared to DWDM systems due to their simpler optical components and lower channel counts. CLI commands such as "estimate-CWDM-deployment-cost" are used to estimate the cost of deploying a CWDM system, taking into account factors such as equipment costs, installation expenses, and ongoing maintenance. However, as data traffic continues to grow and demand for higher capacity increases, the economies of scale may favor DWDM systems over time, particularly in high-capacity backbone networks and data center interconnects.

In summary, both CWDM and DWDM systems play critical roles in modern optical communication networks, offering distinct advantages and applications depending on the specific requirements of the deployment. CLI commands are essential tools for deploying and managing WDM systems, enabling network administrators to configure, optimize, and monitor system performance in real-time. By understanding the differences between CWDM and DWDM systems and their respective strengths and limitations, organizations can make informed decisions when planning and deploying optical networking infrastructure.

Chapter 4: Fiber Optic Sensors and Measurement Techniques

Fiber optic sensing is a groundbreaking technology that has transformed the landscape of sensing and monitoring applications across various industries. CLI commands such as "install-fiber-optic-sensor" are pivotal in deploying fiber optic sensing systems, enabling network administrators to configure sensor parameters, calibrate sensor units, and monitor sensor data in real-time. At its core, fiber optic sensing relies on the principles of light propagation within optical fibers to detect and measure changes in environmental conditions, mechanical strain, temperature variations, and other physical parameters.

One of the fundamental principles underlying fiber optic sensing is the interaction of light with the external environment. CLI commands such as "calibrate-optical-fiber-sensor" are used to calibrate fiber optic sensors, adjusting parameters such as sensitivity, resolution, and dynamic range to suit specific sensing applications. Fiber optic sensors utilize various mechanisms to modulate the intensity, phase, polarization, or wavelength of light in response to changes in the sensed parameter, providing a robust and reliable means of measurement.

Moreover, fiber optic sensors operate based on the principles of light propagation within optical fibers, leveraging the properties of total internal reflection to guide light along the fiber core. CLI commands such as "analyze-optical-fiber-propagation" are used to analyze light propagation characteristics in optical fibers, assessing factors such as attenuation, dispersion, and modal distribution. By exploiting the unique optical properties of optical fibers, fiber optic sensors can achieve high sensitivity, immunity to electromagnetic interference, and multiplexing capabilities for simultaneous measurement of multiple parameters.

Furthermore, fiber optic sensors employ various sensing mechanisms to detect changes in the external environment. CLI commands such as "configure-strain-sensing" are used to configure strain sensing parameters, specifying parameters such as gauge length, sensitivity, and sampling rate. Fiber optic strain sensors utilize the principle of optical phase modulation induced by mechanical strain to measure strain variations along the length of the optical fiber. By monitoring changes in the phase of light propagating through the fiber, strain sensors can accurately measure mechanical deformation and structural integrity in civil engineering, aerospace, and geotechnical applications.

Additionally, fiber optic sensors exploit the principles of light absorption and scattering to measure temperature variations. CLI commands such as "set-temperature-sensing" are used to set temperature sensing parameters, specifying parameters such as wavelength range, resolution, and accuracy. Fiber optic temperature sensors utilize the temperature-dependent absorption or scattering properties of optical fibers or sensing materials to detect changes in temperature. By monitoring changes in the intensity or wavelength of light transmitted through the fiber, temperature sensors can achieve high accuracy and stability over a wide temperature range, making them ideal for applications such as industrial process monitoring, medical diagnostics, and environmental monitoring.

Furthermore, fiber optic sensors can leverage the principles of interferometry to achieve high sensitivity and resolution in displacement and vibration measurements. CLI commands such as "optimize-interferometric-sensing" are used to optimize interferometric sensing parameters, adjusting parameters such as reference arm length, optical path difference, and phase modulation. Interferometric fiber optic sensors utilize the interference patterns generated by combining the light from the sensing and reference arms to detect minute changes in phase or wavelength caused by displacement or vibration. By analyzing the interference fringes, interferometric sensors can achieve sub-nanometer resolution and picometer-level sensitivity, making them ideal for precision metrology, structural health monitoring, and seismic sensing applications.

Moreover, fiber optic sensors can exploit the principles of evanescent field interaction to achieve biochemical sensing and detection. CLI commands such as "configure-biochemical-sensing" are used to configure biochemical sensing parameters, specifying parameters such as sensing probe design, immobilization method, and analyte concentration range. Fiber optic biochemical sensors utilize the evanescent field generated by total internal reflection at the fiber core-cladding interface to interact with biomolecules or chemical analytes present in the surrounding medium. By monitoring changes in the evanescent field intensity or wavelength caused by biomolecular binding events, biochemical sensors can achieve high specificity, sensitivity, and multiplexing capabilities for applications such as medical diagnostics, environmental monitoring, and food safety.

Furthermore, fiber optic sensors can utilize the principles of distributed sensing to achieve spatially continuous monitoring along the length of the optical fiber. CLI commands such as "deploy-distributed-sensing" are used to deploy distributed sensing systems, specifying parameters such as spatial resolution, sampling interval, and measurement range. Distributed fiber optic sensors utilize techniques such as optical time-domain reflectometry (OTDR) or optical frequency domain reflectometry (OFDR) to analyze backscattered light along the optical fiber and detect changes in the scattering properties induced by external stimuli. By analyzing the temporal or spectral characteristics of backscattered light, distributed sensors can achieve high spatial resolution and coverage over long distances, enabling applications such as structural health monitoring, perimeter security, and oil and gas pipeline monitoring.

In summary, fiber optic sensing is a versatile and powerful technology that leverages the principles of light propagation within optical fibers to enable accurate, reliable, and real-time measurement of physical parameters in diverse environments. CLI commands play a crucial role in deploying and managing fiber optic sensing systems, providing administrators with the tools they need to configure, calibrate, and monitor sensor performance. By understanding the underlying principles of fiber optic sensing and its diverse sensing mechanisms, organizations can harness the full potential of this technology to address a wide range of sensing and monitoring challenges across various industries.

The utilization of fiber optic sensing technology in industrial and environmental monitoring has significantly advanced the capabilities of monitoring and managing critical processes and environmental parameters across various sectors. CLI commands such as "deploy-industrial-monitoring" are crucial in deploying fiber optic sensing systems for industrial applications, enabling administrators to configure sensor arrays, calibrate sensor units, and monitor real-time data. These applications encompass a wide range of industries, including manufacturing, energy production, transportation, infrastructure, and environmental monitoring, where fiber optic sensing offers unparalleled advantages in terms of accuracy, reliability, and versatility.

In the manufacturing sector, fiber optic sensing plays a vital role in process monitoring, quality control, and predictive maintenance applications. CLI commands such as "configure-process-monitoring" are used to configure fiber optic sensors for monitoring parameters such as temperature, strain, pressure, and vibration in manufacturing processes. Fiber optic sensors can be integrated into production machinery, pipelines, and structural components to monitor process variables in real-time, ensuring optimal performance, minimizing downtime, and preventing costly equipment failures. For example, in the automotive industry, fiber optic sensors are used to monitor temperature variations in welding processes, ensuring consistent weld quality and structural integrity.

Moreover, fiber optic sensing technology finds extensive applications in the energy sector, particularly in oil and gas production, power generation, and renewable energy systems. CLI commands such as "deploy-oil-and-gas-monitoring" are used to deploy fiber optic sensing systems for monitoring parameters such as temperature, pressure, flow, and chemical composition in oil and gas wells, pipelines, and refineries. Fiber optic sensors can withstand harsh operating conditions and provide accurate and reliable measurements in high-pressure, high-temperature environments, enabling operators to optimize production processes, detect leaks and spills, and ensure safety and environmental compliance.

Additionally, fiber optic sensing technology is instrumental in enhancing the safety and efficiency of transportation systems, including railways, highways, bridges, and tunnels. CLI commands such as "install-structural-monitoring" are used to install fiber optic sensors for monitoring structural health, deformation, and environmental conditions in transportation infrastructure. Fiber optic sensors can detect structural defects, monitor load distribution, and assess the impact of environmental factors such as temperature, humidity, and corrosion on infrastructure integrity. For example, in bridge monitoring applications, fiber optic sensors are embedded in concrete structures to detect cracks, monitor deflection, and assess the effects of traffic loads and environmental conditions on bridge performance.

Furthermore, fiber optic sensing technology is deployed in environmental monitoring applications to monitor air and water quality, detect pollution, and assess the impact of human activities on natural ecosystems. CLI commands such as "deploy-environmental-monitoring" are used to deploy fiber optic sensors for measuring parameters such as temperature, pH, dissolved oxygen, turbidity, and chemical contaminants in air and water bodies. Fiber optic sensors offer high sensitivity and accuracy, enabling continuous monitoring of environmental parameters in remote or inaccessible locations, such as rivers, lakes, oceans, and industrial sites. For example, in water quality monitoring applications, fiber optic sensors are deployed in aquaculture facilities to monitor water temperature, oxygen levels, and nutrient concentrations to ensure optimal conditions for fish health and growth.

Moreover, fiber optic sensing technology is increasingly utilized in smart agriculture applications to monitor soil moisture, temperature, and nutrient levels, optimize irrigation, and enhance crop yield and quality. CLI commands such as "configure-agricultural-monitoring" are used to configure fiber optic sensors for monitoring soil and environmental conditions in agricultural fields.

Fiber optic sensors can be deployed in soil probes, irrigation systems, and crop canopies to provide real-time data on soil moisture content, temperature variations, and plant health parameters. By integrating fiber optic sensing technology into precision agriculture systems, farmers can optimize resource management, reduce water consumption, and improve crop productivity and sustainability.

Additionally, fiber optic sensing technology plays a crucial role in structural health monitoring (SHM) applications to monitor the condition, performance, and integrity of civil infrastructure, including buildings, bridges, dams, and tunnels. CLI commands such as "install-SHM-system" are used to install fiber optic sensors for monitoring structural deformations, vibrations, and environmental conditions in critical infrastructure. Fiber optic sensors can detect changes in strain, displacement, and temperature caused by structural loads, environmental factors, and external events such as earthquakes or blasts. By continuously monitoring structural health parameters, fiber optic sensing systems enable early detection of defects, identification of potential failure modes, and timely intervention to prevent catastrophic failures and ensure public safety.

Furthermore, fiber optic sensing technology is employed in security and surveillance applications to monitor perimeter security, detect intrusions, and identify potential threats in critical infrastructure, military installations, and high-security facilities. CLI commands such as "deploy-security-monitoring" are used to deploy fiber optic sensors for monitoring parameters such as acoustic signals, seismic vibrations, and temperature changes associated with unauthorized activities. Fiber optic sensors can be buried underground, installed along fences or walls, or deployed in sensitive areas to provide real-time detection and notification of security breaches, unauthorized access, or suspicious behavior. By integrating fiber optic sensing technology into security systems, operators can enhance situational awareness, improve response times, and mitigate security risks effectively.

Moreover, fiber optic sensing technology is applied in geophysical monitoring applications to monitor seismic activity, detect subsurface changes, and assess geological hazards such as landslides, earthquakes, and volcanic eruptions. CLI commands such as "configure-geophysical-monitoring" are used to configure fiber optic sensors for detecting acoustic signals, ground vibrations, and temperature changes associated with geological events. Fiber optic sensors can be deployed in boreholes, arrays, or distributed networks to monitor seismic waves, ground deformation, and temperature variations in real-time, providing valuable data for seismic hazard assessment, early warning systems, and disaster mitigation efforts.

In summary, fiber optic sensing technology offers a wide range of applications in industrial and environmental monitoring, enabling accurate, reliable, and real-time measurement of physical parameters across various sectors. CLI commands are essential tools for deploying and managing fiber optic sensing systems, providing administrators with the flexibility to configure, calibrate, and monitor sensor performance according to specific application requirements. By leveraging the capabilities of fiber optic sensing technology, organizations can enhance operational efficiency, ensure safety and compliance, and mitigate risks in critical processes and environmental systems.

Chapter 5: Dense Wavelength Division Multiplexing (DWDM)

DWDM (Dense Wavelength Division Multiplexing) system architecture represents a sophisticated framework designed to facilitate the simultaneous transmission of multiple optical signals over a single optical fiber. CLI commands such as "create-DWDM-network" are essential in deploying DWDM systems, enabling network engineers to configure network elements, allocate wavelengths, and manage optical channels effectively. The architecture of a DWDM system comprises several key components, including transmitters, receivers, optical amplifiers, multiplexers, demultiplexers, and wavelength routers, each playing a critical role in the transmission, amplification, routing, and monitoring of optical signals within the network.

At the core of a DWDM system lies the transmitter unit, responsible for converting electrical signals into optical signals that can be transmitted over the optical fiber. CLI commands such as "configure-DWDM-transmitter" are used to configure transmitter parameters such as modulation format, data rate, and optical power levels. Transmitters employ laser diodes or semiconductor optical amplifiers (SOAs) to generate optical signals at specific wavelengths corresponding to the allocated channels in the DWDM grid. By modulating the intensity, phase, or polarization of the optical signal, transmitters encode data onto the optical carrier for transmission across the network.

Moreover, DWDM systems utilize optical amplifiers to compensate for signal attenuation and maintain signal integrity over long transmission distances. CLI commands such as "deploy-optical-amplifiers" are used to deploy optical amplifiers along the optical fiber to boost the power levels of optical signals. Optical amplifiers, such as erbium-doped fiber amplifiers (EDFAs) or semiconductor optical amplifiers (SOAs), amplify the optical signal without the need for conversion to the electrical domain, enabling high-speed, long-haul transmission with minimal signal degradation. By strategically placing optical amplifiers at regular intervals along the fiber span, DWDM systems can achieve extended reach and higher signal-to-noise ratios, ensuring reliable transmission over thousands of kilometers.

Furthermore, DWDM systems incorporate multiplexers and demultiplexers to combine and separate multiple optical signals at different wavelengths. CLI commands such as "configure-DWDM-multiplexer" and "configure-DWDM-demultiplexer" are used to configure multiplexing and demultiplexing parameters, specifying parameters such as channel spacing, bandwidth, and port assignments. Multiplexers combine individual optical signals from different transmitters into a single composite signal for transmission over the optical fiber, while demultiplexers separate the composite signal into its constituent wavelengths for detection by the receivers. By multiplexing multiple channels onto a single fiber, DWDM systems can achieve high data rates and efficient utilization of the optical spectrum.

Additionally, DWDM systems employ wavelength routers or switches to dynamically route optical signals to different destinations within the network. CLI commands such as "configure-wavelength-router" are used to configure routing tables and switching paths in wavelength router devices. Wavelength routers use optical switches or wavelength-selective switches (WSS) to steer optical signals along different paths based on their wavelength, allowing network operators to establish flexible and reconfigurable optical connections between nodes. By dynamically adjusting the routing of optical signals, wavelength routers enable efficient resource utilization, load balancing, and network optimization in DWDM systems.

Moreover, DWDM systems integrate optical dispersion compensation modules to mitigate chromatic dispersion and polarization mode dispersion effects that can degrade signal quality and limit transmission distances. CLI commands such as "configure-dispersion-compensation" are used to configure dispersion compensation parameters, specifying parameters such as dispersion compensation ratio and compensation distance. Dispersion compensation modules employ fiber Bragg gratings (FBGs), dispersion-compensating fibers (DCF), or chirped fiber gratings (CFGs) to compensate for signal dispersion and maintain signal integrity, enabling high-speed transmission over long-haul fiber optic links.

Furthermore, DWDM systems incorporate optical supervisory channels (OSC) for network management, monitoring, and control functions. CLI commands such as "configure-supervisory-channel" are used to configure OSC parameters, specifying parameters such as wavelength, modulation format, and data rate. OSC channels allow network operators to monitor the performance of DWDM systems, perform fault detection and isolation, and implement dynamic wavelength provisioning and rerouting. By separating management and control traffic from user data traffic, OSC channels enable efficient network management and troubleshooting while minimizing interference with user data transmission.

Additionally, DWDM systems may include optical add-drop multiplexers (OADMs) or reconfigurable optical add-drop multiplexers (ROADMs) to add or drop specific wavelengths at intermediate network nodes without disrupting the transmission of other wavelengths. CLI commands such as "configure-OADM" are used to configure OADM parameters, specifying parameters such as add/drop wavelengths, channel spacing, and passband characteristics. OADMs and ROADM enable network operators to add or remove optical channels at intermediate points in the network, allowing for flexible wavelength routing, wavelength grooming, and wavelength provisioning in DWDM systems.

Moreover, DWDM systems incorporate optical monitoring and management capabilities to ensure optimal performance and reliability. CLI commands such as "monitor-DWDM-performance" are used to monitor key performance metrics such as optical power levels, bit error rates (BER), and signal-to-noise ratios (SNR) in DWDM systems. Optical performance monitoring (OPM) enables network operators to detect signal degradation, identify potential issues, and optimize system parameters to maximize network efficiency and reliability. Additionally, network management systems (NMS) provide centralized control and monitoring of DWDM networks, allowing administrators to configure network elements, provision services, and troubleshoot network problems remotely.

In summary, DWDM system architecture represents a sophisticated framework designed to enable high-capacity, long-haul transmission of optical signals over optical fiber networks. CLI commands play a crucial role in deploying and managing DWDM systems, providing network operators with the tools they need to configure, monitor, and optimize system performance. By understanding the key components and principles of DWDM system architecture, organizations can design and deploy robust and scalable optical communication networks to meet the growing demands for high-speed data transmission and network connectivity.

The deployment of Dense Wavelength Division Multiplexing (DWDM) technology offers numerous benefits and presents several challenges for telecommunications and data networking infrastructure. CLI commands such as "deploy-DWDM-network" are pivotal in deploying DWDM systems, allowing network engineers to configure network elements, allocate wavelengths, and manage optical channels efficiently. DWDM technology enables the transmission of multiple data streams over a single optical fiber by using different wavelengths of light, offering significant advantages in terms of increased bandwidth, scalability, flexibility, and cost-effectiveness. One of the primary benefits of DWDM deployment is its ability to significantly increase the bandwidth capacity of optical fiber networks. By multiplexing multiple data streams onto different wavelengths of light, DWDM systems can achieve aggregate data rates ranging from several terabits per second to hundreds of terabits per second. CLI commands such as "configure-DWDM-transmitter" are used to configure the data rate and modulation format of each optical channel, allowing network operators to optimize the utilization of available bandwidth and meet the growing demand for high-speed data transmission.

Moreover, DWDM deployment enables the scalability and flexibility of optical communication networks, allowing for seamless expansion and upgradeability to accommodate future growth and technological advancements. CLI commands such as "expand-DWDM-network" are used to add new optical channels, upgrade existing network elements, and reconfigure wavelength assignments to support additional services and increased capacity. DWDM systems offer modular and plug-and-play architectures, enabling network operators to deploy new wavelengths, add new network nodes, and integrate emerging technologies without disrupting existing services or infrastructure.

Furthermore, DWDM deployment offers cost-effective solutions for increasing network capacity and meeting the growing demand for bandwidth-intensive applications and services. By leveraging existing fiber optic infrastructure and maximizing the utilization of optical spectrum resources, DWDM systems enable efficient use of network resources and reduce the need for costly fiber optic cable installations. CLI commands such as "optimize-DWDM-network" are used to monitor network performance, optimize wavelength allocations, and minimize operational expenses, ensuring cost-effective deployment and operation of DWDM systems.

Additionally, DWDM deployment enhances the reliability and resilience of optical communication networks by providing built-in redundancy, fault tolerance, and disaster recovery capabilities. CLI commands such as "configure-optical-redundancy" are used to configure network redundancy schemes, such as dual-fiber or ring topologies, to ensure uninterrupted operation and high availability of network services. DWDM systems support automatic protection switching (APS) and optical line protection (OLP) mechanisms, enabling rapid detection and recovery from fiber cuts, equipment failures, or network disruptions.

However, despite its numerous benefits, DWDM deployment also presents several challenges and considerations that must be addressed to ensure successful implementation and operation. One of the key challenges is the complexity of DWDM network planning, design, and optimization, which requires specialized expertise, tools, and resources. CLI commands such as "plan-DWDM-network" are used to perform network modeling, simulation, and analysis to optimize the placement of network elements, minimize signal degradation, and maximize network performance.

Moreover, DWDM deployment requires careful management of optical power levels, signal-to-noise ratios, and chromatic dispersion effects to maintain signal integrity and minimize transmission impairments. CLI commands such as "optimize-optical-power" and "compensate-chromatic-dispersion" are used to adjust optical power levels, optimize amplifier gain profiles, and apply dispersion compensation techniques to mitigate signal degradation and ensure reliable transmission over long-haul fiber optic links.

Furthermore, DWDM deployment may encounter challenges related to interoperability, compatibility, and standardization of network equipment and protocols from different vendors. CLI commands such as "verify-interoperability" are used to test and validate the compatibility of network elements, interfaces, and protocols to ensure seamless integration and interoperability within the DWDM network. Standardization bodies such as the International Telecommunication Union (ITU) play a critical role in defining common standards and specifications for DWDM systems to promote interoperability and vendor neutrality.

Additionally, DWDM deployment requires careful consideration of network security, privacy, and regulatory compliance requirements to protect sensitive data and ensure compliance with industry regulations and legal frameworks. CLI commands such as "implement-network-security" are used to configure access controls, encryption algorithms, and intrusion detection systems to safeguard network infrastructure and prevent unauthorized access or data breaches. Compliance with regulations such as the General Data Protection Regulation (GDPR) and the Health Insurance Portability and Accountability Act (HIPAA) is essential to ensure data privacy and security in DWDM deployments.

In summary, the deployment of DWDM technology offers significant benefits in terms of increased bandwidth, scalability, flexibility, and cost-effectiveness for optical communication networks. However, successful DWDM deployment requires careful planning, design, optimization, and management to address challenges related to complexity, signal integrity, interoperability, security, and regulatory compliance. By leveraging CLI commands and best practices, network operators can overcome these challenges and realize the full potential of DWDM technology to meet the growing demand for high-speed data transmission and network connectivity in the digital age.

Chapter 6: Coherent Optical Communications

Coherent detection and demodulation represent advanced techniques employed in optical communication systems to enhance signal detection sensitivity, improve spectral efficiency, and mitigate impairments such as chromatic dispersion and polarization mode dispersion. CLI commands such as "configure-coherent-receiver" are instrumental in deploying coherent detection systems, allowing network engineers to configure receiver parameters, such as local oscillator frequency and phase, polarization tracking, and signal processing algorithms, to optimize system performance. Coherent detection involves the simultaneous detection of both the amplitude and phase of the received optical signal, enabling the recovery of transmitted data with high fidelity and reliability.

At the heart of coherent detection systems lies the coherent receiver, which consists of a local oscillator (LO) laser, a photodetector, and digital signal processing (DSP) electronics. CLI commands such as "tune-local-oscillator-frequency" are used to adjust the frequency and phase of the LO laser to match the frequency and phase of the received optical signal. The LO laser generates a coherent reference signal that is combined with the received optical signal at the photodetector, allowing for coherent mixing and detection of the optical field. By detecting both the in-phase (I) and quadrature (Q) components of the received signal, coherent receivers enable the extraction of phase and amplitude information, facilitating advanced modulation formats and signal processing techniques.

Moreover, coherent detection systems leverage digital signal processing (DSP) algorithms to demodulate and decode the received optical signal, extracting transmitted data with high accuracy and efficiency. CLI commands such as "configure-DSP-algorithms" are used to specify signal processing parameters, such as modulation format, symbol rate, equalization settings, and forward error correction (FEC) algorithms, to optimize receiver performance. DSP algorithms perform complex mathematical operations, including digital mixing, phase estimation, polarization tracking, and symbol recovery, to demodulate the received signal and recover the transmitted data payload.

Furthermore, coherent detection systems offer several advantages over direct detection techniques, including improved receiver sensitivity, enhanced spectral efficiency, and increased transmission reach. CLI commands such as "optimize-coherent-receiver-settings" are used to adjust receiver parameters, such as optical power levels, polarization states, and modulation formats, to maximize system performance and achieve optimal signal-to-noise ratios (SNR). Coherent detection enables the detection of weak optical signals below the shot noise limit, allowing for longer transmission distances and higher data rates compared to direct detection techniques.

Additionally, coherent detection systems are capable of mitigating impairments such as chromatic dispersion and polarization mode dispersion that can degrade signal quality and limit transmission distances in optical fiber networks. CLI commands such as "compensate-chromatic-dispersion" and "track-polarization-state" are used to apply digital compensation techniques to mitigate dispersion and polarization effects in the received signal. Digital dispersion compensation algorithms adjust the phase and amplitude of the received signal to compensate for chromatic dispersion, while polarization tracking algorithms dynamically adjust receiver settings to maintain optimal polarization alignment and minimize polarization mode dispersion.

Moreover, coherent detection systems enable the implementation of advanced modulation formats and modulation schemes, such as quadrature amplitude modulation (QAM), phase-shift keying (PSK), and differential phase-shift keying (DPSK), to maximize spectral efficiency and increase data throughput. CLI commands such as "configure-modulation-format" are used to specify the modulation format and symbol constellation used in the transmitted signal. Coherent receivers can demodulate complex modulation formats with high spectral efficiency and robustness against optical impairments, enabling the transmission of multiple bits per symbol and achieving higher data rates over optical fiber links.

Furthermore, coherent detection systems support coherent optical transmission techniques, such as coherent optical OFDM (CO-OFDM) and coherent polarization multiplexing (CPM), which further enhance spectral efficiency and transmission capacity in optical communication networks. CLI commands such as "enable-CO-OFDM" and "configure-CPM" are used to deploy and configure advanced transmission techniques in coherent optical systems. CO-OFDM utilizes orthogonal frequency-division multiplexing (OFDM) modulation combined with coherent detection to achieve high spectral efficiency and resistance to channel impairments, while CPM utilizes polarization multiplexing to transmit multiple data streams over orthogonal polarization states, effectively doubling the transmission capacity of optical fiber links.

In summary, coherent detection and demodulation represent essential techniques in optical communication systems for achieving high-performance, high-capacity transmission over optical fiber networks. CLI commands play a crucial role in deploying and configuring coherent detection systems, allowing network operators to optimize receiver performance, mitigate impairments, and maximize spectral efficiency. By leveraging coherent detection techniques and advanced signal processing algorithms, optical communication networks can achieve unprecedented levels of performance, reliability, and efficiency to meet the ever-increasing demand for high-speed data transmission and network connectivity.

Coherent transmission systems represent a cornerstone in modern optical communication networks, enabling high-capacity, long-haul transmission of data over optical fiber infrastructure. CLI commands such as "deploy-coherent-transmission-system" are indispensable for deploying coherent transmission systems, allowing network engineers to configure system parameters, allocate optical channels, and optimize transmission settings to achieve optimal performance. Coherent transmission systems leverage advanced modulation formats, coherent detection techniques, and digital signal processing (DSP) algorithms to maximize spectral efficiency, enhance transmission reach, and mitigate impairments in optical fiber links.

At the core of coherent transmission systems lies the coherent transmitter, which generates optical signals with precise frequency, phase, and amplitude characteristics for transmission over the optical fiber. CLI commands such as "configure-coherent-transmitter" are used to adjust transmitter settings, such as modulation format, symbol rate, and optical power levels, to optimize signal quality and spectral efficiency. Coherent transmitters employ advanced modulation schemes, such as quadrature amplitude modulation (QAM), phase-shift keying (PSK), or frequency-shift keying (FSK), to encode data onto the optical carrier, allowing for high-speed data transmission with improved spectral efficiency and tolerance to optical impairments.

Moreover, coherent transmission systems utilize coherent receivers equipped with local oscillators (LO), photodetectors, and DSP electronics to detect and demodulate the received optical signal. CLI commands such as "configure-coherent-receiver" are used to configure receiver parameters, such as LO frequency, phase tracking, polarization compensation, and DSP algorithms, to optimize signal detection and recovery. Coherent receivers employ heterodyne or homodyne detection techniques to mix the received optical signal with the LO reference signal, enabling coherent detection of both the amplitude and phase of the optical field and extraction of transmitted data with high fidelity and reliability.

Furthermore, coherent transmission systems employ digital signal processing (DSP) algorithms to compensate for impairments such as chromatic dispersion, polarization mode dispersion, and nonlinearities in optical fiber links. CLI commands such as "optimize-DSP-algorithms" are used to adjust DSP parameters, such as equalization settings, phase correction, and forward error correction (FEC) algorithms, to mitigate transmission impairments and improve system performance. DSP algorithms perform complex mathematical operations, including digital filtering, adaptive equalization, and error correction coding, to enhance signal quality, maximize transmission reach, and increase the tolerance to noise and distortion in optical fiber channels.

Additionally, coherent transmission systems support advanced modulation formats and transmission techniques, such as coherent optical OFDM (CO-OFDM), dual-polarization quadrature phase-shift keying (DP-QPSK), and probabilistic constellation shaping (PCS), to further enhance spectral efficiency and transmission capacity. CLI commands such as "enable-CO-OFDM" and "configure-DP-QPSK" are used to deploy and configure advanced modulation formats in coherent transmission systems. CO-OFDM combines orthogonal frequency-division multiplexing (OFDM) modulation with coherent detection to achieve high spectral efficiency and resistance to impairments, while DP-QPSK utilizes dual-polarization modulation to double the transmission capacity of optical fiber links.

Moreover, coherent transmission systems incorporate optical amplification techniques, such as erbium-doped fiber amplifiers (EDFAs) or Raman amplifiers, to compensate for signal attenuation and maintain signal integrity over long-haul transmission distances. CLI commands such as "deploy-optical-amplifiers" are used to install optical amplifiers along the fiber optic link to boost the power levels of optical signals. Optical amplifiers amplify the optical signal without the need for conversion to the electrical domain, enabling high-speed, long-haul transmission with minimal signal degradation and improved receiver sensitivity.

Furthermore, coherent transmission systems support flexible network architectures, such as wavelength-division multiplexing (WDM) and optical mesh networks, to enable dynamic provisioning, rerouting, and restoration of optical connections. CLI commands such as "configure-WDM-network" and "deploy-optical-mesh" are used to configure network elements and establish optical connections between network nodes. Coherent transmission systems enable efficient utilization of optical spectrum resources, dynamic allocation of wavelengths, and intelligent routing of optical traffic, allowing network operators to optimize network capacity, reliability, and performance.

In summary, coherent transmission systems represent a fundamental technology for achieving high-capacity, long-haul transmission of data over optical fiber networks. CLI commands play a critical role in deploying and configuring coherent transmission systems, enabling network operators to optimize system performance, mitigate impairments, and maximize spectral efficiency. By leveraging advanced modulation formats, coherent detection techniques, and digital signal processing algorithms, coherent transmission systems enable reliable, high-speed data transmission over optical fiber links, supporting the growing demand for bandwidth-intensive applications and services in modern telecommunications and data networking environments.

Chapter 7: Fiber Optic Network Security

In the realm of optical networking, ensuring the security and privacy of transmitted data is paramount. CLI commands, such as "configure-encryption" or "enable-encryption," play a crucial role in deploying encryption techniques within optical networks, allowing network administrators to safeguard sensitive information from unauthorized access or interception. Encryption techniques involve the transformation of plaintext data into ciphertext using cryptographic algorithms, making it indecipherable to unauthorized parties. CLI commands are essential for configuring encryption parameters, managing cryptographic keys, and enforcing security policies to protect data confidentiality and integrity in optical communication networks.

One of the fundamental encryption techniques employed in optical networks is symmetric-key encryption, where the same secret key is used for both encryption and decryption of data. CLI commands, such as "generate-symmetric-key" or "set-encryption-key," are utilized to generate cryptographic keys and configure encryption algorithms, such as Advanced Encryption Standard (AES) or Data Encryption Standard (DES), to secure data transmission. Symmetric-key encryption offers high-speed encryption and decryption capabilities, making it suitable for real-time applications and high-speed optical communication links. However, managing and distributing secret keys securely to authorized parties pose significant challenges in symmetric-key encryption.

To address the key management challenges associated with symmetric-key encryption, optical networks often employ asymmetric-key encryption techniques, also known as public-key encryption. CLI commands such as "generate-asymmetric-keys" or "configure-public-key-cryptography" are used to generate key pairs consisting of a public key and a private key, which are mathematically related but cannot be derived from one another. Public-key encryption enables secure key exchange and communication between parties without the need for pre-shared secret keys. However, asymmetric-key encryption algorithms are computationally intensive and may not be suitable for high-speed optical communication systems.

In addition to symmetric-key and asymmetric-key encryption techniques, optical networks may utilize hybrid encryption schemes that combine the advantages of both symmetric and asymmetric encryption. CLI commands such as "configure-hybrid-encryption" or "set-hybrid-encryption-parameters" are employed to configure hybrid encryption algorithms, which involve encrypting data using a symmetric key and then encrypting the symmetric key itself using the recipient's public key. Hybrid encryption offers the efficiency of symmetric-key encryption for bulk data encryption and the security of asymmetric-key encryption for secure key exchange, making it well-suited for securing data transmission in optical networks.

Furthermore, optical networks may implement encryption protocols and standards to ensure interoperability, compatibility, and compliance with industry security guidelines and regulations. CLI commands such as "configure-encryption-protocol" or "set-security-standards" are used to specify encryption protocols, such as Transport Layer Security (TLS) or Internet Protocol Security (IPsec), for securing data communication between network elements and endpoints. Encryption protocols define the procedures for establishing secure communication channels, negotiating cryptographic parameters, and authenticating communicating parties in optical networks.

Moreover, optical networks may employ encryption techniques to protect data at different layers of the network stack, including the physical layer, data link layer, network layer, and application layer. CLI commands such as "enable-encryption-at-physical-layer" or "configure-encryption-at-application-layer" are used to enforce encryption policies and secure data transmission at various network layers. Encrypting data at the physical layer ensures end-to-end security and confidentiality of transmitted data, while encrypting data at higher network layers provides additional security features, such as authentication, integrity protection, and access control.

Additionally, optical networks may deploy encryption techniques to address specific security requirements and use cases, such as secure remote access, virtual private networks (VPNs), secure cloud connectivity, and secure multicast communication. CLI commands such as "configure-VPN-encryption" or "enable-secure-cloud-connectivity" are utilized to implement encryption solutions tailored to specific network architectures and deployment scenarios. Encryption techniques can be applied to protect data in transit, data at rest, and data in use, ensuring comprehensive security and privacy in optical communication networks.

Furthermore, optical networks may integrate encryption capabilities into network elements, such as optical switches, routers, and transceivers, to provide end-to-end encryption and seamless integration with existing network infrastructure. CLI commands such as "configure-encryption-on-network-device" or "enable-encryption-on-optical-transceiver" are used to enable encryption features and functionalities on network devices and optical equipment. Integrated encryption solutions offer centralized management, policy enforcement, and monitoring capabilities, facilitating the deployment and operation of secure optical networks.

In summary, encryption techniques play a critical role in securing data transmission and protecting sensitive information in optical communication networks. CLI commands are essential for deploying encryption solutions, configuring cryptographic parameters, and enforcing security policies to safeguard data confidentiality, integrity, and authenticity in optical networks. By leveraging encryption techniques and standards, network administrators can ensure compliance with security regulations, mitigate security risks, and establish trust in optical communication infrastructure, enabling secure and reliable data communication in today's interconnected world.

Intrusion detection and prevention systems (IDPS) are critical components of cybersecurity infrastructure, aimed at safeguarding fiber optic networks from unauthorized access, malicious activities, and cyber threats. CLI commands, such as "configure-intrusion-detection" or "enable-intrusion-prevention," are instrumental in deploying IDPS within fiber optic networks, allowing network administrators to monitor network traffic, detect suspicious behavior, and take proactive measures to mitigate security risks. IDPS leverage a combination of signature-based detection, anomaly detection, and behavioral analysis techniques to identify and prevent security breaches, intrusions, and cyber attacks in real-time.

One of the primary functions of IDPS in fiber optic networks is to monitor network traffic and analyze data packets for signs of malicious activity or abnormal behavior. CLI commands such as "capture-network-traffic" or "analyze-packet-headers" are used to capture and inspect network packets at various points within the network infrastructure, including routers, switches, and intrusion detection sensors. IDPS examine packet headers, payload contents, and communication patterns to detect known attack signatures, such as denial-of-service (DoS) attacks, port scans, and malware infections, using signature-based detection techniques.

Moreover, IDPS employ anomaly detection techniques to identify deviations from normal network behavior and detect emerging threats or zero-day attacks that evade signature-based detection mechanisms. CLI commands such as "configure-anomaly-detection" or "enable-behavioral-analysis" are used to configure anomaly detection algorithms and statistical models to analyze network traffic patterns, protocol usage, and user behavior for signs of suspicious activity. Anomaly detection techniques leverage machine learning, artificial intelligence, and pattern recognition algorithms to identify abnormal traffic patterns, unauthorized access attempts, and unusual network events indicative of potential security breaches.

Furthermore, IDPS utilize behavioral analysis techniques to establish baselines of normal network behavior and identify deviations or anomalies that may indicate security threats or unauthorized activities. CLI commands such as "establish-behavioral-baselines" or "monitor-user-activity" are used to monitor user interactions, application usage, and system access patterns to detect unauthorized activities, insider threats, and data exfiltration attempts. Behavioral analysis techniques employ heuristic algorithms, anomaly scoring mechanisms, and correlation engines to detect suspicious behavior patterns and trigger alerts or automated response actions.

Additionally, IDPS integrate threat intelligence feeds, vulnerability databases, and security information and event management (SIEM) systems to enhance threat detection capabilities and improve response effectiveness. CLI commands such as "update-threat-intelligence-feeds" or "integrate-with-SIEM" are used to synchronize IDPS with external threat intelligence sources, such as commercial threat feeds, open-source threat databases, and government cybersecurity advisories. Threat intelligence integration enables IDPS to identify known threats, vulnerabilities, and exploits and prioritize security alerts based on the severity and relevance of detected threats.

Moreover, IDPS employ active response mechanisms, such as intrusion prevention systems (IPS) and firewall rules, to block or mitigate detected security threats and prevent unauthorized access to network resources. CLI commands such as "configure-IPS-rules" or "block-suspicious-traffic" are used to deploy IPS policies, firewall rules, and access control lists (ACLs) to block malicious traffic, quarantine infected devices, and enforce security policies within the network infrastructure. Active response mechanisms enable IDPS to take immediate action against detected threats, reducing the risk of data breaches, network downtime, and service disruptions.

Furthermore, IDPS provide real-time alerting and notification mechanisms to alert network administrators and security personnel about detected security incidents, anomalies, or suspicious activities. CLI commands such as "configure-alerting-rules" or "send-alert-notifications" are used to specify alerting thresholds, notification preferences, and escalation procedures for handling security events. IDPS generate alerts, logs, and reports detailing detected threats, security events, and incident response actions, enabling timely incident detection, analysis, and resolution.

Additionally, IDPS support continuous monitoring, auditing, and compliance reporting to ensure adherence to regulatory requirements, industry standards, and organizational security policies. CLI commands such as "perform-security-audits" or "generate-compliance-reports" are used to conduct security audits, vulnerability assessments, and compliance checks to assess the security posture of fiber optic networks. IDPS provide visibility into network security risks, vulnerabilities, and compliance gaps, enabling organizations to address security deficiencies, implement corrective measures, and demonstrate regulatory compliance to stakeholders and regulatory authorities.

In summary, intrusion detection and prevention systems play a crucial role in protecting fiber optic networks from cybersecurity threats, intrusions, and unauthorized access. CLI commands are essential for deploying, configuring, and managing IDPS within fiber optic networks, enabling network administrators to monitor network traffic, detect security incidents, and respond to security threats in real-time. By leveraging signature-based detection, anomaly detection, behavioral analysis, and active response mechanisms, IDPS enhance the security posture of fiber optic networks, safeguarding critical data, applications, and infrastructure from cyber attacks and security breaches.

Chapter 8: Optical Switching Technologies

Electro-optical switches play a pivotal role in modern optical networks, facilitating dynamic routing, wavelength switching, and optical signal manipulation. CLI commands, such as "configure-electro-optical-switch" or "deploy-switching-policies," are integral for deploying electro-optical switches within optical network infrastructures, enabling network operators to manage network resources, optimize traffic routing, and enhance network flexibility. Electro-optical switches utilize electro-optical modulation techniques to control the transmission of optical signals and redirect them along different paths within the network, enabling efficient utilization of network resources and enabling on-demand provisioning of optical connections.

At the heart of electro-optical switches lie optical modulators, which modulate the intensity, phase, or polarization of optical signals in response to electrical control signals. CLI commands, such as "configure-optical-modulator" or "set-modulation-parameters," are utilized to configure modulation parameters, bias voltages, and modulation formats for optical modulators within electro-optical switches. Optical modulators employ various modulation schemes, such as amplitude modulation (AM), phase modulation (PM), or polarization modulation (PolM), to control the transmission of optical signals and achieve efficient optical switching and routing.

Moreover, electro-optical switches employ optical couplers, splitters, or beam combiners to route optical signals between different input and output ports within the switch. CLI commands, such as "configure-optical-routing" or "connect-input-output-ports," are used to configure routing paths, optical connections, and switching configurations in electro-optical switches. Optical routing algorithms determine the optimal path for routing optical signals based on factors such as signal quality, available bandwidth, and network topology, ensuring efficient and reliable transmission of optical data across the network.

Furthermore, electro-optical switches support various switching architectures, including space-switching, time-switching, wavelength-switching, and polarization-switching, to accommodate diverse network requirements and traffic patterns. CLI commands, such as "select-switching-architecture" or "configure-switching-modes," are employed to specify switching modes, scheduling policies, and switching parameters for electro-optical switches. Space-switching architectures route optical signals based on physical connections between input and output ports, while time-switching architectures utilize time-division multiplexing (TDM) to allocate time slots for signal transmission. Wavelength-switching architectures leverage wavelength-division multiplexing (WDM) to route optical signals based on their wavelengths, while polarization-switching architectures utilize polarization diversity to separate and route optical signals based on their polarization states.

Additionally, electro-optical switches support reconfigurable optical add-drop multiplexing (ROADM) functionalities, enabling dynamic provisioning and management of optical channels within optical networks. CLI commands, such as "configure-ROADM-functionality" or "enable-optical-channel-routing," are used to deploy ROADM capabilities and configure optical channel add-drop operations within electro-optical switches. ROADM-enabled switches allow network operators to add, drop, or reroute optical channels at intermediate network nodes, enabling flexible wavelength routing, wavelength grooming, and wavelength routing optimization in optical transport networks.

Moreover, electro-optical switches integrate with network management systems, control plane protocols, and software-defined networking (SDN) frameworks to enable centralized control, programmability, and automation of optical network resources. CLI commands, such as "integrate-with-NMS" or "configure-SDN-interface," are utilized to integrate electro-optical switches with network management systems and SDN controllers, enabling centralized monitoring, provisioning, and orchestration of optical network resources. SDN-based control plane protocols, such as OpenFlow, NETCONF, or RESTCONF, facilitate dynamic control and configuration of electro-optical switches, enabling network-wide optimization and traffic engineering.

Furthermore, electro-optical switches support optical cross-connect (OXC) functionalities, allowing flexible and efficient interconnection of optical fibers and network elements within optical transport networks. CLI commands, such as "configure-OXC-matrix" or "set-cross-connect-routing," are employed to deploy OXC capabilities and configure optical cross-connect operations within electro-optical switches. OXC-enabled switches enable non-blocking, all-optical switching of optical signals at the wavelength level, facilitating seamless network scalability, wavelength grooming, and service provisioning in optical transport networks.

In summary, electro-optical switches are essential building blocks in optical communication networks, enabling dynamic routing, wavelength switching, and optical signal manipulation. CLI commands play a vital role in deploying, configuring, and managing electro-optical switches, enabling network operators to optimize network performance, enhance resource utilization, and meet evolving network demands. By leveraging electro-optical switches, network operators can achieve efficient and flexible optical connectivity, support diverse traffic patterns, and deliver high-performance optical communication services in today's demanding networking environments.

All-optical switching techniques represent a significant advancement in optical networking, offering the capability to manipulate optical signals without converting them into electrical form. CLI commands such as "configure-all-optical-switch" or "deploy-optical-switching-technique" are instrumental in deploying these techniques, enabling network operators to achieve ultra-fast, low-latency, and energy-efficient optical signal routing and switching. All-optical switching techniques exploit nonlinear optical effects, optical interference phenomena, and photonic components to realize optical signal processing functions, including routing, switching, wavelength conversion, and signal regeneration, entirely in the optical domain.

One of the fundamental principles underlying all-optical switching techniques is the exploitation of nonlinear optical effects, such as four-wave mixing (FWM), cross-phase modulation (XPM), and self-phase modulation (SPM), to induce changes in the optical properties of the transmission medium. CLI commands, such as "enable-nonlinear-optical-effects" or "configure-optical-phase-modulation," are utilized to configure optical components and parameters to exploit these nonlinear effects for signal processing applications. Nonlinear effects allow optical signals to interact with each other or with the medium itself, enabling functionalities such as wavelength conversion, signal regeneration, and optical signal amplification, all while preserving the optical nature of the signals.

Furthermore, all-optical switching techniques leverage optical interferometers, such as Mach-Zehnder interferometers (MZIs) or Michelson interferometers, to achieve optical signal routing and switching based on the interference of optical waves. CLI commands such as "configure-optical-interferometer" or "set-interference-routing-policy" are used to deploy interferometric-based switching techniques, enabling network operators to manipulate the phase, amplitude, or polarization of optical signals to control their routing paths. Interferometric-based switches exploit the constructive and destructive interference of optical waves to steer signals along different paths or routes within the optical network, enabling efficient and flexible optical signal routing without the need for electrical conversion.

Moreover, all-optical switching techniques utilize semiconductor optical amplifiers (SOAs), optical resonators, or nonlinear optical fibers to achieve gain and amplification of optical signals entirely in the optical domain. CLI commands such as "configure-optical-amplification" or "set-amplification-gain" are employed to configure amplification parameters and gain profiles for optical amplifiers within all-optical switches. Optical amplifiers enhance the strength and quality of optical signals without converting them into electrical form, enabling long-haul transmission, signal regeneration, and wavelength amplification in optical communication systems.

Additionally, all-optical switching techniques exploit optical logic gates, photonic integrated circuits (PICs), or optical cross-connects (OXCs) to perform complex signal processing and routing operations directly in the optical domain. CLI commands such as "configure-optical-logic-gate" or "deploy-photonic-integrated-circuit" are used to deploy these photonic components and configure their operational parameters for specific signal processing tasks. Optical logic gates enable logical operations, such as AND, OR, and NOT, to be performed on optical signals, allowing for sophisticated signal processing and decision-making capabilities within optical networks.

Furthermore, all-optical switching techniques support wavelength-selective switching, allowing network operators to route optical signals based on their wavelengths or colors. CLI commands such as "configure-wavelength-selective-switching" or "set-wavelength-routing-policy" are employed to configure wavelength-selective switches and specify routing policies based on optical wavelengths. Wavelength-selective switches utilize techniques such as diffraction gratings, arrayed waveguide gratings (AWGs), or tunable optical filters to route optical signals to specific output ports based on their wavelengths, enabling wavelength-division multiplexing (WDM) and wavelength-division demultiplexing (WDDM) functionalities in optical networks.

Moreover, all-optical switching techniques support polarization-based switching, allowing network operators to manipulate the polarization states of optical signals to control their routing paths. CLI commands such as "configure-polarization-selective-switching" or "set-polarization-routing-policy" are utilized to configure polarization-selective switches and define routing policies based on optical polarization states. Polarization-selective switches exploit polarization diversity techniques, such as polarization beam splitters (PBS) or polarization-maintaining fibers (PMFs), to route optical signals along different polarization axes, enabling polarization-division multiplexing (PDM) and polarization-based signal processing in optical networks.

In summary, all-optical switching techniques offer unprecedented capabilities for manipulating optical signals entirely in the optical domain, enabling ultra-fast, low-latency, and energy-efficient signal processing and routing in optical communication systems. CLI commands play a crucial role in deploying and configuring these techniques, enabling network operators to achieve flexible, scalable, and high-performance optical networks capable of meeting the ever-increasing demands of modern telecommunications and data communications applications.

Chapter 9: Fiber Optic Network Management and Optimization

Remote fiber monitoring and management are critical aspects of modern optical network operations, ensuring the reliability, performance, and security of fiber optic infrastructure across diverse deployment scenarios. CLI commands such as "enable-remote-monitoring" or "configure-management-policies" are indispensable for deploying these techniques, allowing network operators to remotely monitor fiber optic links, detect potential faults or anomalies, and proactively manage network resources to prevent service disruptions and optimize network performance. Remote fiber monitoring and management encompass a range of technologies, tools, and protocols designed to facilitate real-time monitoring, fault detection, performance analysis, and configuration management of fiber optic networks from a centralized location.

One of the key components of remote fiber monitoring and management is optical time-domain reflectometry (OTDR), which enables network operators to perform non-intrusive testing of fiber optic cables and identify the location and severity of fiber faults or impairments. CLI commands such as "run-OTDR-test" or "analyze-OTDR-results" are utilized to initiate OTDR tests, collect measurement data, and analyze reflectometry traces to assess fiber quality and detect potential issues such as breaks, bends, or splice losses along the fiber span. OTDR testing provides valuable insights into fiber health, enabling proactive maintenance and troubleshooting to minimize service downtime and optimize network availability.

Moreover, remote fiber monitoring and management employ optical performance monitoring (OPM) techniques to continuously monitor key performance parameters of optical signals, such as power levels, signal-to-noise ratio (SNR), and chromatic dispersion (CD), to ensure optimal network performance and reliability. CLI commands such as "enable-OPM" or "monitor-optical-performance" are employed to configure OPM parameters, collect performance data, and generate performance reports for analysis and troubleshooting. OPM enables network operators to identify degradation trends, anticipate potential network issues, and take corrective actions to maintain service quality and meet service level agreements (SLAs) with customers.

Furthermore, remote fiber monitoring and management utilize distributed fiber optic sensing (DFOS) technologies, such as distributed temperature sensing (DTS) and distributed acoustic sensing (DAS), to monitor environmental conditions and detect physical disturbances or intrusions along fiber optic cables in real time. CLI commands such as "deploy-DFOS-sensors" or "monitor-environmental-conditions" are employed to install DFOS sensors along fiber routes, collect sensing data, and analyze sensor outputs to detect abnormal events or security breaches. DFOS enables network operators to enhance physical security, detect unauthorized access, and protect critical infrastructure assets, such as pipelines, railways, and perimeter fences, in remote or inaccessible locations.

Additionally, remote fiber monitoring and management leverage network management systems (NMS) and supervisory control and data acquisition (SCADA) systems to provide centralized visibility, control, and automation of fiber optic network operations. CLI commands such as "integrate-with-NMS" or "configure-SCADA-interface" are used to integrate fiber monitoring and management systems with NMS and SCADA platforms, enabling unified management, alarm handling, and performance analysis across the entire network infrastructure. NMS and SCADA systems enable network operators to monitor network health, configure network devices, and respond to alarms or events in real time, enhancing operational efficiency and reducing downtime.

Moreover, remote fiber monitoring and management incorporate machine learning (ML) and artificial intelligence (AI) algorithms to analyze large volumes of monitoring data, identify patterns or anomalies, and predict potential network failures or performance degradations before they occur. CLI commands such as "train-ML-model" or "deploy-AI-analytics" are employed to develop ML models and deploy AI-based analytics platforms for predictive maintenance, anomaly detection, and root cause analysis in fiber optic networks. ML and AI technologies enable proactive network management, optimize resource allocation, and improve network reliability and resilience in dynamic and complex operational environments.

Furthermore, remote fiber monitoring and management support secure remote access and authentication mechanisms, such as virtual private networks (VPNs), secure shell (SSH), and Transport Layer Security (TLS), to ensure the confidentiality, integrity, and availability of monitoring and management data transmitted over the network. CLI commands such as "establish-VPN-tunnel" or "configure-SSH-access" are utilized to set up secure communication channels and enforce access control policies for remote management operations. Secure remote access mechanisms enable authorized personnel to remotely configure network devices, perform maintenance tasks, and troubleshoot network issues without compromising network security or exposing sensitive data to unauthorized access.

In summary, remote fiber monitoring and management play a crucial role in ensuring the reliability, performance, and security of fiber optic networks in today's interconnected and data-driven world. CLI commands are essential for deploying, configuring, and managing remote monitoring and management systems, enabling network operators to proactively monitor network health, detect potential issues, and respond promptly to ensure uninterrupted service delivery and meet the evolving demands of digital communication and connectivity. By leveraging advanced monitoring technologies, centralized management platforms, and secure remote access mechanisms, organizations can optimize network operations, enhance service quality, and deliver superior user experiences in the era of digital transformation and connectivity.

Performance optimization strategies are essential for ensuring the efficient operation and maximal utilization of computing resources in various technological domains. CLI commands such as "optimize-performance" or "tune-system-settings" are pivotal for deploying these techniques, allowing administrators to fine-tune system configurations, enhance application performance, and maximize resource efficiency to meet the performance requirements of critical workloads and applications. Performance optimization encompasses a wide range of techniques, including hardware optimization, software optimization, workload balancing, and resource allocation, aimed at improving system responsiveness, throughput, and scalability while minimizing latency, bottlenecks, and resource contention.

One of the fundamental aspects of performance optimization is hardware tuning, which involves configuring hardware components such as CPUs, memory modules, storage devices, and network interfaces to maximize their performance and efficiency. CLI commands such as "set-CPU-clock-speed" or "configure-memory-timings" are utilized to adjust hardware parameters and optimize their operation for specific workloads and applications. Hardware tuning techniques include overclocking CPUs, enabling hardware acceleration features, optimizing memory access patterns, and configuring storage devices with appropriate RAID levels and caching mechanisms to improve data access and transfer rates.

Moreover, software optimization plays a crucial role in performance enhancement, involving the optimization of application code, algorithms, data structures, and system software to minimize resource utilization and improve execution efficiency. CLI commands such as "compile-with-optimizations" or "profile-application-performance" are used to compile, profile, and optimize software applications for better performance. Software optimization techniques include code refactoring, loop unrolling, instruction scheduling, and memory optimization to reduce CPU cycles, minimize cache misses, and improve overall program efficiency and responsiveness.

Furthermore, workload balancing strategies aim to distribute computing tasks and resources evenly across system components, such as CPU cores, memory channels, and storage devices, to avoid resource contention and maximize throughput. CLI commands such as "configure-load-balancer" or "monitor-resource-utilization" are employed to balance workloads dynamically and ensure optimal resource allocation in multi-core processors, distributed storage systems, and virtualized environments. Workload balancing techniques include task scheduling algorithms, job prioritization policies, and affinity settings to optimize resource utilization, minimize response times, and improve system throughput and scalability.

Additionally, caching mechanisms are commonly used to optimize performance by storing frequently accessed data or instructions in high-speed cache memory closer to the CPU, reducing latency and improving data access times. CLI commands such as "configure-cache-policy" or "flush-cache-entries" are utilized to manage cache configurations and eviction policies for optimal performance. Caching techniques include instruction caching, data caching, and pre-fetching mechanisms to minimize memory latency, reduce disk I/O operations, and enhance application responsiveness and throughput.

Moreover, network optimization strategies focus on optimizing network traffic, reducing latency, and improving throughput in communication networks by optimizing network protocols, packet processing, and transmission parameters. CLI commands such as "tune-network-parameters" or "optimize-Quality-of-Service" are used to configure network settings and parameters to achieve optimal performance. Network optimization techniques include congestion control algorithms, traffic shaping policies, and Quality of Service (QoS) mechanisms to prioritize traffic, minimize packet loss, and ensure consistent network performance for critical applications and services.

Furthermore, database optimization techniques aim to improve database performance, scalability, and reliability by optimizing database schema, queries, indexing, and transaction management to minimize response times and maximize throughput. CLI commands such as "optimize-database-schema" or "analyze-query-execution-plans" are employed to optimize database configurations and query execution plans for better performance. Database optimization techniques include index tuning, query optimization, data partitioning, and database caching to reduce disk I/O operations, improve query execution times, and enhance overall database performance and scalability.

In summary, performance optimization strategies are essential for maximizing the efficiency and responsiveness of computing systems and applications in diverse technological environments. CLI commands play a critical role in deploying and configuring performance optimization techniques, enabling administrators to fine-tune system parameters, optimize hardware and software configurations, balance workloads, and optimize network and database performance for optimal operation. By implementing effective performance optimization strategies, organizations can improve system responsiveness, enhance user experience, and meet the performance requirements of mission-critical applications and services in today's demanding and competitive technological landscape.

Chapter 10: Next-Generation Fiber Optic Networks

Photonic Integrated Circuits (PICs) represent a cutting-edge technology that has revolutionized the field of photonics by enabling the integration of multiple optical components and functions onto a single chip, analogous to electronic integrated circuits. PICs have emerged as a key enabler for a wide range of applications in telecommunications, data communications, sensing, imaging, and quantum computing, offering significant advantages in terms of size, weight, power consumption, and performance compared to traditional discrete optical components. CLI commands such as "design-PIC-layout" or "simulate-PIC-performance" are instrumental in the design, simulation, and fabrication of PICs, allowing engineers to optimize device performance, functionality, and manufacturability for specific application requirements.

One of the primary advantages of PICs is their compact footprint and high level of integration, which enables the realization of complex optical functions and systems on a small chip, leading to miniaturization, cost reduction, and improved system reliability. CLI commands such as "synthesize-PIC-design" or "optimize-chip-layout" are used to synthesize and optimize PIC designs, integrating various optical components such as lasers, modulators, detectors, waveguides, and filters on a single substrate or chip. PICs enable the implementation of sophisticated functionalities, such as wavelength division multiplexing

(WDM), optical signal processing, and quantum photonic circuits, in a compact and scalable platform, facilitating the development of advanced photonic systems for diverse applications.

Moreover, PICs offer superior performance characteristics, including low insertion loss, high spectral purity, wide bandwidth, and low crosstalk, compared to discrete optical components, owing to the precise control and integration of optical elements on a monolithic substrate. CLI commands such as "analyze-PIC-performance" or "optimize-device-characteristics" are employed to analyze and optimize PIC performance metrics, such as insertion loss, extinction ratio, bandwidth, and signal-to-noise ratio (SNR), to meet specific application requirements. PICs enable the realization of high-performance optical transceivers, switches, amplifiers, and sensors with enhanced functionality and reliability, facilitating the deployment of advanced photonic systems in telecommunications networks, data centers, and sensor networks.

Furthermore, PICs enable the implementation of reconfigurable and programmable optical devices and circuits, offering dynamic control and flexibility in tailoring device functionality and performance to adapt to changing network conditions and application requirements. CLI commands such as "program-PIC-functionality" or "reconfigure-device-settings" are utilized to program or reconfigure PICs to perform specific optical functions, such as wavelength tuning, polarization control, and optical switching, in real time. Reconfigurable PICs facilitate the development of agile and adaptive photonic systems capable of dynamic reconfiguration and optimization, enhancing network flexibility, resilience, and efficiency in response to evolving traffic demands and service requirements.

Additionally, PICs enable the integration of advanced functionalities, such as quantum photonics, nonlinear optics, and on-chip signal processing, paving the way for the development of next-generation photonic technologies for quantum communication, quantum computing, and nonlinear optical signal processing applications. CLI commands such as "integrate-quantum-components" or "simulate-nonlinear-effects" are employed to integrate and simulate advanced functionalities in PICs, enabling the exploration of novel photonic phenomena and the development of innovative photonic devices and systems with enhanced capabilities. PICs offer a platform for exploring and exploiting quantum effects, such as entanglement, superposition, and teleportation, for secure communication, ultrafast computing, and quantum information processing, opening up new avenues for scientific research and technological innovation.

Moreover, PICs facilitate the integration of photonics with complementary metal-oxide-semiconductor (CMOS) electronics, enabling the realization of hybrid photonic-electronic integrated circuits for enhanced system performance and functionality. CLI commands such as "fabricate-hybrid-PIC" or "test-PIC-electronic-interaction" are utilized to fabricate and characterize hybrid PICs, integrating photonics and electronics on a single chip for applications such as on-chip optical interconnects, optoelectronic sensing, and neuromorphic computing. Hybrid PICs leverage the strengths of both photonics and electronics, combining the high-speed, low-loss transmission capabilities of optics with the processing power and integration density of electronics, to enable new paradigms in information processing, communication, and sensing.

In summary, Photonic Integrated Circuits (PICs) represent a transformative technology that has revolutionized the field of photonics, enabling the integration of multiple optical components and functions onto a single chip for diverse applications in telecommunications, sensing, imaging, and quantum computing. CLI commands play a crucial role in the design, simulation, fabrication, and characterization of PICs, enabling engineers to optimize device performance, functionality, and manufacturability for specific application requirements. With their compact size, high level of integration, superior performance characteristics, and versatility, PICs are poised to drive innovation and advancement in photonic systems and applications, paving the way for future breakthroughs in communication, computation, and sensing technologies.

Quantum communication networks represent a groundbreaking paradigm in information exchange, leveraging the principles of quantum mechanics to enable secure, high-speed, and reliable communication over long distances. CLI commands such as "configure-quantum-network" or "deploy-quantum-communication-protocol" are integral to deploying and managing quantum communication networks, allowing administrators to configure network settings, establish quantum key distribution (QKD) protocols, and monitor network performance. Quantum communication networks exploit the unique properties of quantum systems, such as superposition, entanglement, and uncertainty, to offer unprecedented levels of security, privacy, and efficiency in transmitting and processing information.

One of the fundamental concepts in quantum communication networks is quantum key distribution (QKD), which enables the generation and distribution of cryptographic keys between distant parties with unconditional security based on the principles of quantum mechanics. CLI commands such as "initialize-QKD-protocol" or "establish-quantum-key-exchange" are used to initiate and configure QKD protocols, allowing users to generate shared secret keys encoded in quantum states and use them for secure communication. QKD protocols, such as BB84 and E91, exploit the properties of quantum states, such as photon polarization or quantum entanglement, to detect eavesdropping attempts and ensure the secrecy and integrity of transmitted data, making them immune to interception or decryption by classical adversaries.

Moreover, quantum communication networks leverage quantum entanglement, a phenomenon where quantum particles become correlated in such a way that the state of one particle is instantaneously correlated with the state of another, regardless of the distance between them. CLI commands such as "create-entangled-pair" or "measure-quantum-correlation" are utilized to generate and manipulate entangled states for quantum communication applications, enabling the implementation of quantum teleportation, superdense coding, and distributed quantum computing protocols. Entanglement-based communication protocols offer advantages such as ultra-secure key distribution, quantum teleportation of qubits, and enhanced channel capacity, enabling novel applications in quantum cryptography, quantum teleportation, and distributed quantum information processing.

Furthermore, quantum communication networks exploit quantum repeaters and amplifiers to extend the range and improve the reliability of quantum communication over long distances. CLI commands such as "deploy-quantum-repeater" or "configure-quantum-amplification" are employed to deploy and configure quantum repeater stations and amplification techniques, allowing the distribution of entangled states and quantum information over large-scale quantum networks. Quantum repeaters utilize entanglement swapping and purification techniques to overcome the inherent limitations of quantum communication, such as photon loss and decoherence, enabling the establishment of entanglement and quantum communication links over intercontinental distances.

Additionally, quantum communication networks enable secure quantum teleportation, a process that allows the transfer of quantum states from one location to another without physical transmission of the qubits themselves. CLI commands such as "initiate-quantum-teleportation" or "verify-quantum-state-transfer" are used to initiate and verify quantum teleportation protocols, enabling the secure transmission of quantum information between remote quantum nodes. Quantum teleportation protocols rely on entanglement and classical communication to transfer the state of a qubit from a sender to a receiver, ensuring the secure transmission of quantum information without the risk of interception or tampering, making them ideal for secure quantum communication and quantum information processing applications.

Moreover, quantum communication networks offer unique advantages in terms of information security and privacy, as they are inherently resistant to eavesdropping and interception due to the principles of quantum mechanics. CLI commands such as "monitor-quantum-channel-security" or "detect-eavesdropping-attempts" are employed to monitor the security of quantum communication channels and detect any unauthorized access or tampering attempts. Quantum communication protocols, such as QKD and quantum key distribution, leverage the principles of quantum mechanics, such as the no-cloning theorem and the uncertainty principle, to ensure the secrecy and integrity of transmitted data, offering unparalleled levels of security and privacy in communication networks.

In summary, quantum communication networks represent a revolutionary approach to secure, high-speed, and reliable communication, leveraging the principles of quantum mechanics to enable secure key distribution, quantum teleportation, and quantum information processing over long distances. CLI commands play a crucial role in deploying and managing quantum communication networks, enabling administrators to configure network settings, establish secure communication protocols, and monitor network performance. With their unique advantages in terms of security, privacy, and efficiency, quantum communication networks hold immense promise for revolutionizing communication, cryptography, and information processing in the digital age.

BOOK 3
FIBER OPTIC INFRASTRUCTURE DESIGN AND
IMPLEMENTATION
PRACTICAL STRATEGIES FOR PROFESSIONALS

ROB BOTWRIGHT

Chapter 1: Understanding Infrastructure Requirements

Assessing bandwidth needs and scalability is a critical aspect of designing and managing modern communication networks, ensuring that they can meet current and future demands for data transmission and processing. CLI commands such as "measure-bandwidth-requirements" or "analyze-network-traffic" are essential for assessing the bandwidth needs of a network, allowing administrators to monitor traffic patterns, identify bottlenecks, and predict future capacity requirements. By analyzing network traffic using tools like Wireshark or SNMP (Simple Network Management Protocol), administrators can gain insights into data consumption patterns, application usage, and peak traffic periods, enabling them to allocate resources efficiently and plan for network expansion or upgrades. Furthermore, assessing the scalability of a network involves evaluating its ability to accommodate growth in users, devices, and data traffic while maintaining performance and reliability. CLI commands such as "evaluate-network-scalability" or "simulate-traffic-load" are utilized to assess the scalability of a network, allowing administrators to model different scenarios, simulate traffic loads, and analyze performance metrics under varying conditions. Network scalability encompasses factors such as capacity planning, resource provisioning, and network topology design, ensuring that the network can scale seamlessly to support increasing demands without compromising quality of service or user experience.

Moreover, assessing bandwidth needs and scalability requires a comprehensive understanding of network requirements, user expectations, and technological advancements. CLI commands such as "conduct-network-audit" or "perform-capacity-planning" are employed to assess the current state of the network infrastructure, identify potential limitations or constraints, and develop strategies for enhancing scalability and performance. Network audits involve evaluating factors such as network topology, hardware capabilities, software configurations, and traffic patterns to determine areas for improvement and optimization.

Additionally, assessing bandwidth needs involves considering factors such as data volume, application requirements, and user behavior to determine the optimal bandwidth allocation for different network segments and services. CLI commands such as "configure-quality-of-service" or "prioritize-network-traffic" are used to implement quality of service (QoS) policies, traffic shaping, and bandwidth management techniques to ensure that critical applications receive adequate bandwidth and priority over less time-sensitive traffic. QoS mechanisms such as traffic prioritization, traffic shaping, and congestion management enable administrators to allocate bandwidth based on application requirements, service level agreements (SLAs), and user preferences, ensuring optimal performance and user experience across the network.

Furthermore, assessing scalability involves evaluating the capacity of network infrastructure, including switches, routers, servers, and storage systems, to accommodate growth in data traffic, users, and applications. CLI commands such as "scale-network-infrastructure" or "deploy-virtualization-technologies" are employed to scale network resources horizontally or vertically, adding additional capacity, redundancy, or virtualization capabilities to meet growing demands. Horizontal scalability involves adding more resources, such as servers or switches, to distribute workload and increase capacity, while vertical scalability involves upgrading existing hardware or software components to handle larger volumes of data or users.

Moreover, assessing bandwidth needs and scalability requires considering factors such as network architecture, topology, and technology trends to develop a future-proof network infrastructure that can adapt to evolving requirements and emerging technologies. CLI commands such as "evaluate-network-architecture" or "assess-technology-trends" are utilized to analyze network design principles, identify potential bottlenecks, and explore new technologies or standards that can enhance scalability, performance, and efficiency. Network architects and administrators must stay abreast of industry developments, such as cloud computing, virtualization, software-defined networking (SDN), and network function virtualization (NFV), to leverage innovative solutions and best practices for building scalable and resilient networks.

In summary, assessing bandwidth needs and scalability is crucial for designing, managing, and optimizing modern communication networks to meet the growing demands for data transmission and processing. CLI commands play a vital role in analyzing network traffic, evaluating scalability, and implementing optimization strategies to ensure optimal performance, reliability, and efficiency. By adopting a proactive approach to assessing bandwidth needs and scalability, organizations can build agile and resilient network infrastructures capable of supporting current and future requirements, enabling them to stay competitive in the rapidly evolving digital landscape.

When establishing a new network infrastructure or upgrading an existing one, assessing physical space and access requirements is paramount to ensure seamless deployment and efficient operation. CLI commands such as "measure-space-availability" or "analyze-access-requirements" are essential tools in this process, allowing network administrators to gather crucial data and make informed decisions regarding equipment placement, cabling routes, and accessibility considerations. By carefully evaluating the physical environment and access constraints, organizations can optimize resource utilization, minimize deployment challenges, and enhance network reliability and performance.

The first step in determining physical space and access requirements is to conduct a thorough assessment of the site where the network infrastructure will be deployed. CLI commands such as "survey-site-layout" or "inspect-building-blueprints" can be used to gather information about the available space, layout, and infrastructure of the site. This may involve measuring dimensions, identifying potential obstacles or constraints, and assessing environmental factors such as temperature, humidity, and electromagnetic interference (EMI) levels. By conducting a comprehensive site survey, network administrators can gain insights into the physical characteristics of the site and identify any challenges or limitations that may impact the deployment of network equipment.

Once the site assessment is complete, the next step is to determine the specific physical space requirements for housing network equipment, such as servers, switches, routers, and cabling infrastructure. CLI commands such as "calculate-rack-space" or "design-equipment-layout" can aid in determining the amount of rack space, floor space, and clearance needed to accommodate the network infrastructure. Factors to consider include the size and form factor of the equipment, ventilation and cooling requirements, power and cabling needs, and accessibility for maintenance and servicing. By accurately estimating space requirements, organizations can ensure that adequate provisions are made for equipment installation and future expansion, minimizing the risk of overcrowding, overheating, or accessibility issues.

In addition to space considerations, assessing access requirements is essential to ensure that network equipment can be easily accessed for installation, maintenance, and troubleshooting purposes. CLI commands such as "evaluate-access-points" or "plan-cable-routing" can assist in identifying access points, cable pathways, and connectivity requirements for network equipment. This may involve determining the location of entry points, cable conduits, and utility connections, as well as assessing the feasibility of cable routing and management within the physical space. By planning access routes and cable pathways in advance, organizations can streamline installation workflows, minimize cable congestion, and facilitate future maintenance and upgrades with minimal disruption to operations.

Furthermore, it is essential to consider regulatory and compliance requirements when assessing physical space and access for network infrastructure deployment. CLI commands such as "check-regulatory-guidelines" or "ensure-compliance-standards" can help verify adherence to building codes, safety regulations, and industry standards governing network infrastructure installations. This may include requirements related to fire safety, electrical codes, environmental regulations, and accessibility standards for persons with disabilities. By ensuring compliance with relevant regulations and standards, organizations can mitigate legal risks, ensure workplace safety, and maintain operational integrity while deploying network infrastructure.

Moreover, it is crucial to consider scalability and future growth when assessing physical space and access requirements for network infrastructure deployment.

CLI commands such as "plan-scalable-architecture" or "forecast-expansion-needs" can aid in designing a flexible and scalable infrastructure that can accommodate future growth and evolving business needs. This may involve reserving additional space for future equipment expansion, implementing modular or stackable hardware configurations, and planning for easy scalability and upgrades as demand increases. By adopting a scalable infrastructure design, organizations can future-proof their network deployments, minimize costly retrofitting efforts, and adapt quickly to changing business requirements and technological advancements.

In summary, determining physical space and access requirements is a critical aspect of network infrastructure deployment, requiring careful planning, assessment, and consideration of various factors. CLI commands play a vital role in gathering data, analyzing requirements, and making informed decisions regarding equipment placement, cabling routes, and accessibility considerations. By accurately assessing physical space and access requirements, organizations can optimize resource utilization, ensure regulatory compliance, facilitate future scalability, and enhance the reliability and performance of their network infrastructure deployments.

Chapter 2: Site Survey and Planning for Fiber Optic Deployment

Site surveys are essential pre-deployment tasks in network infrastructure planning, involving the assessment of physical locations to gather data crucial for effective deployment. CLI commands such as "conduct-site-survey" or "analyze-environmental-factors" are indispensable tools for this purpose, enabling network administrators to evaluate various factors and make informed decisions about equipment placement, cabling routes, and environmental considerations. Site surveys encompass a range of factors, including physical layout, infrastructure readiness, environmental conditions, and regulatory compliance, all of which play critical roles in ensuring the success of network deployments and minimizing potential issues or disruptions.

One of the primary factors to consider when conducting site surveys is the physical layout of the site, including building dimensions, room configurations, and structural features. CLI commands such as "measure-room-dimensions" or "inspect-building-layout" can assist in gathering data about the layout of the site, identifying potential obstacles or constraints that may affect equipment placement and cable routing. This may involve measuring room dimensions, assessing ceiling heights, identifying load-bearing walls or columns, and determining the location of utility connections, entry points, and access routes. By understanding the physical layout of the site, network administrators can plan equipment placement and cable pathways more effectively, minimizing the risk of interference or obstruction.

Another critical consideration during site surveys is the readiness of existing infrastructure to support network deployments, including power, cooling, and connectivity provisions. CLI commands such as "evaluate-power-capacity" or "check-cooling-systems" can aid in assessing the adequacy of existing infrastructure to meet the requirements of network equipment. This may involve examining electrical outlets, circuit capacities, and backup power sources, as well as evaluating HVAC systems, ventilation, and temperature control mechanisms. Additionally, network administrators must verify the availability of network connectivity, including internet access, LAN connections, and telecommunications infrastructure, to ensure seamless integration with existing networks and services.

Furthermore, environmental conditions play a significant role in site surveys, as they can impact the performance and reliability of network equipment. CLI commands such as "monitor-temperature-humidity" or "measure-electromagnetic-interference" can help assess environmental factors that may affect network deployments. This may include monitoring temperature and humidity levels to ensure optimal operating conditions for network equipment, as well as measuring electromagnetic interference (EMI) levels to identify potential sources of interference that could degrade network performance. Additionally, considerations such as physical security, fire suppression systems, and environmental hazards must be evaluated to ensure the safety and security of network installations.

Moreover, regulatory compliance is a crucial aspect of site surveys, as network deployments must adhere to local building codes, safety regulations, and industry standards. CLI commands such as "verify-regulatory-compliance" or "ensure-code-compliance" can assist in verifying compliance with relevant regulations and standards governing network infrastructure installations. This may involve checking for compliance with fire safety regulations, electrical codes, accessibility standards, and environmental regulations, as well as obtaining necessary permits and approvals from regulatory authorities. By ensuring regulatory compliance, organizations can mitigate legal risks, ensure workplace safety, and maintain operational integrity while deploying network infrastructure.

Additionally, site surveys should consider future scalability and expansion needs, as network deployments must be designed to accommodate future growth and evolving business requirements. CLI commands such as "plan-scalable-infrastructure" or "forecast-expansion-needs" can help in designing a flexible and scalable infrastructure that can adapt to changing demands. This may involve reserving additional space for future equipment expansion, implementing modular or stackable hardware configurations, and planning for easy scalability and upgrades as demand increases. By adopting a scalable infrastructure design, organizations can future-proof their network deployments, minimize costly retrofitting efforts, and adapt quickly to evolving business needs and technological advancements.

In summary, conducting site surveys is a critical aspect of network infrastructure planning, requiring careful assessment of various factors to ensure successful deployments. CLI commands play a vital role in gathering data, analyzing requirements, and making informed decisions about equipment placement, cabling routes, and environmental considerations. By considering factors such as physical layout, infrastructure readiness, environmental conditions, regulatory compliance, and future scalability, organizations can optimize resource utilization, minimize deployment challenges, and enhance the reliability and performance of their network infrastructure deployments.

Planning pathways and routes for cable installation is a crucial aspect of network infrastructure deployment, requiring meticulous consideration of factors such as physical layout, accessibility, environmental conditions, and regulatory compliance. CLI commands such as "map-cable-pathways" or "plan-cable-routing" are indispensable tools for this purpose, enabling network administrators to design efficient and organized pathways for running cables while minimizing potential issues or disruptions. Pathway planning involves determining the optimal routes for cable installation, identifying potential obstacles or constraints, and ensuring compliance with relevant regulations and standards to facilitate smooth and seamless deployments.

The first step in planning cable pathways is to conduct a comprehensive assessment of the physical layout of the site where the network infrastructure will be deployed. CLI commands such as "inspect-site-layout" or "analyze-building-blueprints" can assist in gathering data about the layout of the site, including building dimensions, room configurations, and structural features. This may involve measuring room dimensions, identifying potential obstacles such as walls, columns, or ceilings, and determining the most efficient routes for running cables between equipment locations. By understanding the physical layout of the site, network administrators can identify suitable pathways for cable installation and plan routes that minimize cable lengths and avoid potential obstructions.

Once the site layout has been assessed, the next step is to identify access points and entryways for running cables between different locations within the facility. CLI commands such as "locate-access-points" or "map-entryways" can help in identifying suitable entry points for cable installation and determining the most direct routes for running cables between equipment rooms, workstations, and other network endpoints. This may involve identifying existing cable conduits, utility shafts, or cable trays that can be used to route cables, as well as evaluating the feasibility of installing new pathways or conduits where necessary. By planning access points and entryways in advance, network administrators can streamline cable installation workflows and minimize disruption to operations during deployment.

Additionally, it is essential to consider environmental conditions and regulatory requirements when planning cable pathways and routes. CLI commands such as "assess-environmental-factors" or "verify-regulatory-compliance" can aid in evaluating environmental factors such as temperature, humidity, and electromagnetic interference (EMI), as well as ensuring compliance with relevant regulations and standards governing cable installations. This may include adhering to fire safety regulations, electrical codes, and industry standards for cable routing and management, as well as obtaining necessary permits and approvals from regulatory authorities. By considering environmental and regulatory factors, network administrators can ensure the safety, reliability, and compliance of cable installations while minimizing the risk of damage or interference.

Furthermore, pathway planning should take into account future scalability and expansion needs to accommodate evolving business requirements and technological advancements. CLI commands such as "plan-scalable-routing" or "forecast-expansion-needs" can assist in designing flexible and scalable cable pathways that can adapt to changing demands. This may involve reserving additional capacity in cable conduits or trays for future cable installations, implementing modular or stackable cable management systems, and planning for easy scalability and upgrades as demand increases. By adopting a scalable pathway design, organizations can future-proof their network deployments and minimize the need for costly retrofitting efforts in the future.

Moreover, it is essential to document cable pathways and routes effectively to facilitate maintenance, troubleshooting, and future upgrades. CLI commands such as "document-cable-pathways" or "create-routing-diagrams" can help in documenting the planned pathways and routes for cable installation, including detailed diagrams, schematics, and documentation of cable types, lengths, and termination points. This documentation can serve as a valuable reference for network administrators and technicians during installation and maintenance activities, helping to ensure consistency, accuracy, and efficiency in cable management and troubleshooting efforts.

In summary, planning pathways and routes for cable installation is a critical aspect of network infrastructure deployment, requiring careful consideration of various factors to ensure successful and efficient deployments. CLI commands play a vital role in gathering data, analyzing requirements, and designing organized and scalable cable pathways that minimize disruption to operations and maximize the reliability and performance of network installations. By considering factors such as physical layout, accessibility, environmental conditions, regulatory compliance, and future scalability, organizations can optimize cable installation workflows, minimize deployment challenges, and enhance the reliability and performance of their network infrastructure deployments.

Chapter 3: Fiber Optic Cable Installation Techniques

Direct burial and conduit installation are two common methods for deploying underground cables, each with its advantages, challenges, and considerations. CLI commands such as "deploy-direct-burial-cables" or "install-cables-in-conduit" are instrumental in executing these techniques, enabling network administrators to choose the most suitable method based on factors such as environmental conditions, terrain, and project requirements. Direct burial involves burying cables directly into the ground without protective enclosures, while conduit installation entails placing cables within protective conduits or ducts to shield them from external elements and facilitate future maintenance or upgrades.

Direct burial is often favored for its simplicity and cost-effectiveness, as it eliminates the need for additional materials such as conduits and reduces installation time and labor. CLI commands such as "prepare-cable-routes-for-burial" or "dig-trenches-for-cable-installation" can assist in preparing the ground for direct burial, including excavating trenches, laying cables, and backfilling the trenches to cover the cables securely. This method is commonly used in rural or sparsely populated areas where underground utilities are less likely to be disturbed and where environmental conditions are favorable for direct burial.

However, direct burial may pose challenges in certain environments, such as areas with rocky or hard soil, high groundwater levels, or frequent landscaping activities. CLI commands such as "assess-soil-conditions" or "conduct-environmental-surveys" can help in evaluating soil conditions, drainage patterns, and other environmental factors that may impact direct burial installations. In areas with challenging soil conditions, additional equipment such as trenching machines or rock saws may be required to excavate trenches for cable installation, increasing installation costs and complexity. Moreover, direct burial cables are more susceptible to damage from external factors such as digging, landscaping, or accidental excavation, requiring careful planning and marking to prevent accidental damage.

Conversely, conduit installation provides added protection and flexibility for underground cables, making it suitable for areas with challenging terrain, harsh environmental conditions, or strict regulatory requirements. CLI commands such as "lay-conduit-routes" or "install-cable-ducts" can aid in laying conduits or ducts along planned cable routes, providing a protective enclosure for cables and facilitating future maintenance or upgrades. Conduits can be made of various materials such as PVC, HDPE, or metal, depending on factors such as soil conditions, depth of burial, and environmental hazards.

One of the primary advantages of conduit installation is its ability to protect cables from external elements such as moisture, debris, and physical damage, reducing the risk of cable failures and minimizing downtime. CLI commands such as "seal-conduit-joints" or "install-conduit-protective-caps" can help in ensuring watertight seals and protective measures to safeguard cables within conduits. Additionally, conduits provide flexibility for future upgrades or modifications, as cables can be easily added, removed, or replaced within the conduit without the need for extensive excavation or disruption.

However, conduit installation may involve higher upfront costs and longer installation times compared to direct burial, as it requires additional materials such as conduits, fittings, and installation equipment. CLI commands such as "calculate-conduit-lengths" or "estimate-materials-costs" can assist in estimating the required materials and costs for conduit installations, including conduits, fittings, couplings, and installation accessories. Moreover, conduit installations may require permits or approvals from regulatory authorities, as well as adherence to specific standards and guidelines governing underground utility installations.

In summary, direct burial and conduit installation are two common methods for deploying underground cables, each with its advantages, challenges, and considerations. CLI commands play a crucial role in executing these techniques, enabling network administrators to choose the most suitable method based on factors such as environmental conditions, terrain, and project requirements. By carefully evaluating the pros and cons of each method and considering factors such as cost, complexity, and long-term maintenance, organizations can make informed decisions to ensure the successful and reliable deployment of underground cable infrastructure.

Aerial fiber optic cable installation is a common method for deploying fiber optic cables above ground, typically along utility poles or other elevated structures. CLI commands such as "plan-aerial-cable-routes" or "inspect-existing-utility-poles" play a crucial role in executing these procedures, enabling network administrators to plan, deploy, and maintain aerial fiber optic cable installations efficiently. This method offers several advantages, including cost-effectiveness, ease of access, and reduced environmental impact compared to underground installations.

The first step in aerial fiber optic cable installation is to conduct a thorough assessment of the proposed route and existing infrastructure to identify suitable locations for installing cables and support structures. CLI commands such as "survey-potential-cable-routes" or "analyze-pole-loading-capacity" can aid in evaluating factors such as pole spacing, clearance requirements, and environmental conditions to ensure the safe and efficient deployment of aerial cables. This may involve inspecting existing utility poles, determining pole loading capacities, and assessing clearance requirements for other utilities such as power lines or telecommunications cables.

Once the route has been surveyed and evaluated, the next step is to plan the installation process and obtain necessary permits or approvals from regulatory authorities. CLI commands such as "obtain-permits-for-aerial-installation" or "submit-plans-for-regulatory-review" can assist in navigating the permitting process and ensuring compliance with local regulations and standards governing aerial installations. This may include obtaining right-of-way permits, utility easements, or approvals from property owners or municipal authorities before commencing installation activities.

After obtaining necessary permits and approvals, the installation team can begin deploying aerial fiber optic cables according to the planned route. CLI commands such as "attach-cables-to-utility-poles" or "secure-cable-supports" can help in attaching cables to utility poles or support structures safely and securely. This may involve using cable lashing techniques, brackets, or clamps to secure cables to poles and ensure proper tension and support to prevent sagging or damage.

During the installation process, it is essential to follow industry best practices and safety guidelines to minimize the risk of accidents or injuries. CLI commands such as "conduct-safety-training-for-installation-team" or "implement-fall-protection-measures" can help in ensuring that installation teams are adequately trained and equipped to work safely at heights and in challenging environments. This may include providing personal protective equipment (PPE), such as safety harnesses, helmets, and high-visibility clothing, as well as implementing fall protection measures such as guardrails or safety nets to prevent falls from elevated structures.

Additionally, it is crucial to document the installation process and maintain accurate records of cable routes, attachment points, and support structures for future reference and maintenance. CLI commands such as "document-aerial-cable-installation" or "create-installation-logs" can aid in documenting the installation process and capturing essential information such as cable lengths, attachment methods, and pole locations. This documentation can serve as a valuable reference for troubleshooting, maintenance, and future expansion of the aerial fiber optic network.

Once the installation is complete, the final step is to conduct thorough testing and verification of the installed cables to ensure proper functionality and performance. CLI commands such as "conduct-fiber-optic-testing" or "verify-signal-quality" can assist in testing the integrity and performance of the installed cables, including measuring signal loss, attenuation, and optical power levels. This may involve using specialized testing equipment such as optical time-domain reflectometers (OTDRs), optical power meters, and optical spectrum analyzers to analyze and diagnose any issues or discrepancies in the installed cables.

In summary, aerial fiber optic cable installation procedures involve careful planning, execution, and documentation to ensure the safe and efficient deployment of fiber optic cables above ground. CLI commands play a vital role in facilitating these procedures, enabling network administrators to plan, deploy, and maintain aerial fiber optic cable installations effectively. By following industry best practices, safety guidelines, and regulatory requirements, organizations can ensure the successful deployment of aerial fiber optic networks to support the growing demand for high-speed telecommunications and broadband services.

Chapter 4: Splicing and Termination Best Practices

Fusion splicing is a fundamental method used to join two optical fibers by fusing or welding them together to create a continuous optical path for transmitting light signals. CLI commands such as "prepare-fiber-ends-for-splicing" or "execute-fusion-splice" are essential in deploying fusion splicing techniques, enabling network technicians to achieve reliable and low-loss connections between optical fibers. This technique offers several advantages, including low insertion loss, high mechanical strength, and minimal signal attenuation compared to other splicing methods such as mechanical splicing or adhesive bonding.

The first step in fusion splicing is to prepare the ends of the optical fibers by stripping off the protective coatings and cleaving the fibers to create flat, perpendicular ends for optimal alignment during the splicing process. CLI commands such as "strip-fiber-coatings" or "cleave-fiber-ends" can aid in preparing the fibers for splicing, ensuring clean and precise fiber ends for accurate alignment and fusion. This may involve using specialized fiber stripping tools and cleavers to remove the protective coatings and create smooth, flat surfaces on the fiber ends.

Once the fibers are prepared, they are aligned and fused together using a fusion splicing machine, which applies heat to melt the fiber ends and fuses them together to form a continuous optical connection. CLI commands such as "align-fiber-ends" or "initiate-fusion-splicing" can assist in aligning the fiber ends with sub-micron precision and initiating the fusion splicing process. This may involve using motorized stages and imaging systems to align the fiber ends accurately before applying heat to fuse them together.

During the fusion splicing process, it is essential to control various parameters such as temperature, fusion time, and arc power to ensure consistent and reliable splices with low insertion loss and high mechanical strength. CLI commands such as "adjust-fusion-parameters" or "monitor-fusion-process" can help in controlling and monitoring these parameters in real-time to achieve optimal splice quality. This may involve adjusting fusion parameters such as arc voltage, arc current, fusion time, and electrode pressure based on factors such as fiber type, diameter, and coating material.

After the fusion splicing process is complete, the splice point is typically protected with a splice sleeve or protective enclosure to provide mechanical strength and protect the splice from environmental factors such as moisture, dust, and physical damage. CLI commands such as "install-splice-sleeve" or "encapsulate-splice-point" can aid in installing splice sleeves or protective enclosures over the splice point, ensuring a secure and durable connection between the optical fibers. This may involve using heat-shrink sleeves or mechanical splice enclosures to protect the splice and provide strain relief to prevent fiber breakage or signal loss.

Once the splicing process is finished, the splice point is typically tested and verified using optical testing equipment such as an optical time-domain reflectometer (OTDR) or optical power meter to ensure proper alignment, low insertion loss, and high optical performance. CLI commands such as "conduct-splice-testing" or "verify-splice-quality" can assist in testing and verifying the quality of the splice, including measuring insertion loss, return loss, and reflectance levels to ensure optimal signal transmission through the spliced fibers.

In summary, fusion splicing is a critical technique used to join optical fibers in telecommunications and fiber optic networks, enabling reliable and low-loss connections for transmitting light signals. CLI commands play a crucial role in deploying fusion splicing techniques, facilitating the preparation, alignment, fusion, and testing of optical fibers to achieve high-quality splices with minimal signal loss and optimal optical performance. By following best practices and utilizing advanced fusion splicing equipment and techniques, network technicians can ensure the successful deployment of fusion splices to support the growing demand for high-speed and reliable optical communication systems.

Mechanical splicing is a technique used to join two optical fibers using mechanical alignment and adhesion methods, without the need for heat or fusion. CLI commands such as "prepare-fiber-ends-for-mechanical-splicing" or "execute-mechanical-splice" are integral in deploying mechanical splicing techniques, enabling network technicians to achieve reliable and low-loss connections between optical fibers. This method offers several advantages, including simplicity, ease of deployment, and reusability, making it a cost-effective solution for temporary or field installations where fusion splicing may not be practical.

The process of mechanical splicing begins with preparing the ends of the optical fibers by stripping off the protective coatings and cleaving the fibers to create flat, perpendicular ends for precise alignment and bonding. CLI commands such as "strip-fiber-coatings" or "cleave-fiber-ends" assist in preparing the fibers for mechanical splicing, ensuring clean and smooth fiber ends for accurate alignment and bonding. This may involve using specialized fiber stripping tools and cleavers to remove the protective coatings and create flat surfaces on the fiber ends.

Once the fibers are prepared, they are aligned and bonded together using a mechanical splice assembly, which typically consists of precision-machined alignment sleeves and index matching gel or adhesive to facilitate optical coupling between the fibers. CLI commands such as "assemble-mechanical-splice" or "apply-index-matching-gel" aid in assembling the mechanical splice and applying the necessary bonding materials to secure the fibers in place. This may involve inserting the prepared fiber ends into the alignment sleeves and applying a small amount of index matching gel or adhesive to ensure optical continuity and minimal signal loss at the splice point.

During the mechanical splicing process, it is essential to ensure precise alignment and bonding of the fiber ends to minimize insertion loss and maintain optimal optical performance. CLI commands such as "align-fiber-ends" or "bond-fiber-ends" assist in aligning the fiber ends within the splice assembly and applying the necessary pressure or tension to bond them securely together. This may involve using alignment jigs, microscope inspection, or automated alignment systems to achieve accurate alignment and bonding of the fiber ends.

After the mechanical splicing process is complete, the splice assembly is typically protected with a splice enclosure or protective sleeve to provide mechanical stability and protect the splice from environmental factors such as moisture, dust, and physical damage. CLI commands such as "install-splice-enclosure" or "protect-mechanical-splice" aid in installing splice enclosures or protective sleeves over the splice point, ensuring a secure and durable connection between the optical fibers. This may involve using heat-shrink sleeves, mechanical splice enclosures, or protective caps to encapsulate the splice and provide strain relief to prevent fiber breakage or signal loss.

Once the splicing process is finished, the splice point is typically tested and verified using optical testing equipment such as an optical time-domain reflectometer (OTDR) or optical power meter to ensure proper alignment, low insertion loss, and high optical performance. CLI commands such as "conduct-splice-testing" or "verify-splice-quality" assist in testing and verifying the quality of the splice, including measuring insertion loss, return loss, and reflectance levels to ensure optimal signal transmission through the spliced fibers.

In summary, mechanical splicing is a valuable technique used to join optical fibers in telecommunications and fiber optic networks, providing reliable and low-loss connections without the need for heat or fusion. CLI commands play a crucial role in deploying mechanical splicing techniques, facilitating the preparation, alignment, bonding, and testing of optical fibers to achieve high-quality splices with minimal signal loss and optimal optical performance. By following best practices and utilizing advanced mechanical splicing equipment and techniques, network technicians can ensure the successful deployment of mechanical splices to support the growing demand for high-speed and reliable optical communication systems.

Chapter 5: Fiber Optic Network Design Considerations

Fiber optic distribution architecture refers to the design and arrangement of optical fibers within a network to facilitate the efficient transmission of data between various points. CLI commands such as "create-fiber-optic-distribution-plan" or "configure-fiber-optic-distribution-network" are vital in deploying fiber optic distribution architecture, enabling network administrators to plan, design, and implement robust fiber optic infrastructures tailored to specific requirements and environments. This architecture plays a crucial role in modern telecommunications and data networking systems, providing high-speed, reliable, and scalable connectivity for a wide range of applications and services.

The design of fiber optic distribution architecture begins with assessing the requirements and objectives of the network, including factors such as bandwidth needs, geographic coverage, scalability, and redundancy. CLI commands such as "analyze-network-requirements" or "assess-coverage-needs" aid in gathering and analyzing relevant data to inform the design process, ensuring that the resulting architecture meets the needs of the intended applications and users. This may involve conducting site surveys, traffic analysis, and capacity planning to determine the optimal distribution of optical fibers and network resources.

Once the requirements are understood, the next step is to design the physical layout and topology of the fiber optic distribution network. CLI commands such as "design-fiber-optic-network-topology" or "layout-fiber-optic-distribution-routes" assist in creating detailed plans and diagrams that outline the placement of optical fibers, equipment locations, cable routes, and connectivity options within the network. This may involve selecting appropriate fiber optic cables, splice enclosures, distribution panels, and termination points to ensure efficient signal transmission and easy maintenance.

During the deployment phase, network administrators use CLI commands such as "install-fiber-optic-cabling" or "deploy-fiber-optic-distribution-equipment" to physically install and configure the necessary components of the distribution architecture. This may involve laying fiber optic cables along predefined routes, installing splice enclosures and distribution panels in central or remote locations, and terminating fibers at end-user premises or network equipment. Proper installation techniques and adherence to industry standards are essential to ensure the reliability and performance of the distribution network.

As the distribution architecture is deployed, network administrators use CLI commands such as "test-fiber-optic-connections" or "verify-network-performance" to conduct testing and verification procedures to ensure that the installed components meet specified requirements and standards. This may involve using optical time-domain reflectometers (OTDRs), optical power meters, and other testing equipment to measure signal loss, attenuation, and reflection levels along the fiber optic links. Additionally, network administrators may perform network performance tests to validate throughput, latency, and reliability under various operating conditions.

Once the distribution architecture is operational, ongoing maintenance and management are essential to ensure its continued performance and reliability. CLI commands such as "monitor-fiber-optic-network-health" or "manage-distribution-equipment-configurations" assist in monitoring network status, detecting faults or performance issues, and making necessary adjustments or repairs to maintain optimal operation. This may involve implementing network management systems (NMS) to monitor network traffic, equipment status, and performance metrics in real-time, enabling proactive maintenance and troubleshooting.

In summary, fiber optic distribution architecture plays a critical role in modern telecommunications and data networking systems, providing the foundation for high-speed, reliable, and scalable connectivity. CLI commands are indispensable tools in planning, designing, deploying, and managing fiber optic distribution networks, enabling network administrators to create robust infrastructures that meet the needs of today's demanding applications and services. By following best practices and leveraging advanced technologies, organizations can build and maintain distribution architectures that support their current and future networking requirements.

When designing fiber optic distribution architecture for telecommunications networks, various factors must be considered to ensure optimal performance, scalability, and reliability. CLI commands such as "configure-fiber-distribution" or "deploy-fiber-optic-network" are instrumental in implementing the architecture, allowing network engineers to configure the distribution of fiber optic cables based on the specific requirements of the network.

One of the primary considerations in fiber optic distribution architecture is the choice between centralized and decentralized architectures. In a centralized architecture, all fiber optic cables converge at a central location, such as a data center or main distribution frame (MDF). CLI commands like "set-centralized-distribution" or "configure-MDF-location" can be used to establish this architecture, enabling network administrators to consolidate network resources and simplify management and maintenance tasks.

On the other hand, decentralized architectures distribute fiber optic cables across multiple locations, such as remote facilities or edge data centers. CLI commands such as "deploy-decentralized-network" or "configure-edge-distribution" facilitate the implementation of decentralized architectures, allowing network engineers to extend network connectivity to remote locations and improve redundancy and fault tolerance.

Another crucial consideration in fiber optic distribution architecture is the choice between point-to-point and multipoint architectures. In a point-to-point architecture, each fiber optic cable is dedicated to connecting two endpoints, providing a direct and secure communication link between them. CLI commands like "establish-point-to-point-connection" or "configure-dedicated-links" can be used to deploy point-to-point architectures, allowing network administrators to ensure secure and reliable communication between specific locations.

In contrast, multipoint architectures utilize shared fiber optic cables to connect multiple endpoints, allowing for more efficient use of network resources and cost savings. CLI commands such as "configure-multipoint-connection" or "establish-shared-links" facilitate the implementation of multipoint architectures, enabling network engineers to accommodate the communication needs of multiple endpoints within the network.

When designing fiber optic distribution architecture, it is essential to consider the bandwidth requirements of the network and select appropriate fiber optic cable types and transmission technologies to meet these requirements. CLI commands like "calculate-bandwidth-needs" or "select-optimal-fiber-types" assist in assessing bandwidth needs and choosing the right fiber optic cables, such as single-mode or multi-mode fibers, and transmission technologies, such as wavelength division multiplexing (WDM) or time division multiplexing (TDM).

Furthermore, network engineers must consider factors such as network scalability, redundancy, and future growth when designing fiber optic distribution architecture. CLI commands such as "plan-for-scalability" or "implement-redundancy-strategies" help in designing scalable and resilient network architectures that can accommodate future expansion and withstand unexpected network failures.

In summary, designing fiber optic distribution architecture requires careful consideration of various factors, including centralized vs. decentralized architectures, point-to-point vs. multipoint architectures, bandwidth requirements, scalability, redundancy, and future growth. CLI commands play a crucial role in deploying and managing fiber optic networks, enabling network engineers to configure, monitor, and optimize network resources to meet the communication needs of modern telecommunications networks. By leveraging CLI commands and following best practices in network design, organizations can build robust and reliable fiber optic distribution architectures that support the growing demand for high-speed and high-bandwidth communication services.

Chapter 6: Implementing Redundancy and Fault Tolerance

Implementing redundancy strategies is essential for ensuring high availability and fault tolerance in critical network segments. CLI commands play a pivotal role in deploying these strategies, enabling network administrators to configure redundant paths, devices, and protocols to mitigate the impact of network failures and maintain uninterrupted service.

One common redundancy strategy is the implementation of redundant physical paths for critical network connections. CLI commands like "configure-redundant-path" or "set-up-link-redundancy" can be used to configure redundant fiber optic cables, Ethernet links, or wireless connections between network devices. By establishing multiple physical paths for data transmission, organizations can minimize the risk of network downtime due to cable cuts, equipment failures, or other physical infrastructure issues.

In addition to physical redundancy, network administrators can implement device-level redundancy to enhance fault tolerance in critical network segments. CLI commands such as "configure-device-redundancy" or "set-up-high-availability" facilitate the deployment of redundant network devices, such as routers, switches, or firewalls, to ensure continuous operation in the event of a device failure. This may involve configuring protocols like Virtual Router Redundancy Protocol

(VRRP) or Hot Standby Router Protocol (HSRP) to enable automatic failover between redundant devices.

Another effective redundancy strategy is the implementation of protocol-level redundancy to provide backup communication paths and protocols for critical network segments. CLI commands like "configure-protocol-redundancy" or "set-up-protocol-failover" can be used to configure protocols such as Spanning Tree Protocol (STP), Rapid Spanning Tree Protocol (RSTP), or EtherChannel to create redundant communication paths and automatically switch to backup protocols in the event of a primary path failure.

Furthermore, organizations can implement geographic redundancy by deploying redundant network infrastructure in geographically diverse locations to minimize the impact of localized disasters or outages. CLI commands such as "deploy-geographic-redundancy" or "set-up-disaster-recovery-site" assist in configuring redundant data centers, network hubs, or cloud regions in different geographical regions to ensure business continuity and data resilience.

Network administrators can also leverage network monitoring and management tools to proactively identify and mitigate potential points of failure in critical network segments. CLI commands like "monitor-network-health" or "analyze-fault-reports" enable administrators to monitor network performance metrics, detect anomalies or errors, and take proactive measures to address potential issues before they impact service availability.

In addition to technical redundancy strategies, organizations should also consider implementing redundant power and cooling systems, physical security measures, and backup data storage solutions to further enhance the resilience of critical network segments. CLI commands such as "configure-power-redundancy" or "set-up-backup-storage" support the deployment of redundant infrastructure components to ensure continuous operation in the face of power outages, environmental hazards, or data loss events.

Overall, implementing redundancy strategies for critical network segments is essential for maintaining high availability, fault tolerance, and business continuity in modern network environments. CLI commands empower network administrators to deploy redundant paths, devices, protocols, and infrastructure components to mitigate the risk of network downtime and ensure uninterrupted service delivery to end users. By leveraging CLI commands and following best practices in redundancy design and deployment, organizations can build robust and resilient network architectures that meet the demands of today's mission-critical applications and services.

In modern network environments, fault detection and automatic failover mechanisms are crucial for maintaining high availability and ensuring uninterrupted service delivery. CLI commands are instrumental in deploying these mechanisms, allowing network administrators to configure proactive monitoring, detection, and failover processes to quickly respond to network faults and minimize downtime.

One of the primary mechanisms for fault detection is the implementation of proactive monitoring tools and protocols to continuously monitor the health and performance of network devices and links. CLI commands like "configure-proactive-monitoring" or "set-up-network-monitoring-protocols" enable administrators to deploy monitoring solutions such as Simple Network Management Protocol (SNMP), NetFlow, or packet capture tools to capture and analyze network traffic, identify performance anomalies, and detect potential points of failure.

Additionally, administrators can configure fault detection mechanisms such as Link Layer Discovery Protocol (LLDP) or Cisco Discovery Protocol (CDP) to automatically discover and map network devices and links, facilitating accurate network topology visualization and identification of potential points of failure. CLI commands like "enable-LLDP" or "configure-CDP" support the deployment of these protocols, allowing administrators to monitor network device connectivity and detect changes in device status or connectivity.

Another critical aspect of fault detection is the implementation of proactive health checks and diagnostic tests to identify potential issues before they impact service availability. CLI commands such as "schedule-health-checks" or "run-diagnostic-tests" enable administrators to configure automated tests and checks to assess the health and performance of network devices, interfaces, and services, and generate alerts or notifications in case of abnormalities or failures.

Furthermore, administrators can deploy proactive fault detection mechanisms such as Bidirectional Forwarding Detection (BFD) or Ethernet OAM (Operations, Administration, and Maintenance) to detect link failures or network outages in real-time and trigger automatic failover to redundant paths or devices. CLI commands like "configure-BFD" or "enable-Ethernet-OAM" facilitate the deployment of these mechanisms, allowing administrators to ensure rapid fault detection and failover in the event of network disruptions.

Once a fault is detected, automatic failover mechanisms come into play to seamlessly redirect traffic to redundant paths or devices to maintain uninterrupted service delivery. CLI commands such as "configure-automatic-failover" or "set-up-redundant-paths" enable administrators to configure protocols like Virtual Router Redundancy Protocol (VRRP), Hot Standby Router Protocol (HSRP), or Dynamic Host Configuration Protocol (DHCP) failover to automatically switch traffic to backup devices or paths in the event of a primary path failure.

Moreover, administrators can deploy dynamic routing protocols such as Open Shortest Path First (OSPF) or Border Gateway Protocol (BGP) with fast convergence capabilities to enable rapid rerouting of traffic in the event of network failures. CLI commands like "configure-OSPF" or "enable-BGP-fast-convergence" support the deployment of these protocols, allowing administrators to ensure efficient and reliable failover in dynamic network environments.

In summary, fault detection and automatic failover mechanisms are essential components of modern network architectures, enabling organizations to maintain high availability and ensure uninterrupted service delivery. CLI commands empower network administrators to deploy proactive monitoring, detection, and failover processes, facilitating rapid response to network faults and minimizing downtime. By leveraging CLI commands and implementing best practices in fault detection and automatic failover, organizations can build resilient and reliable network infrastructures that meet the demands of today's mission-critical applications and services.

Chapter 7: Fiber Optic Power Budgeting and Link Budget Analysis

In optical networking, calculating optical power budgets is crucial for ensuring reliable and efficient transmission of data over fiber optic cables. CLI commands play a significant role in deploying these calculations, enabling network administrators to assess the performance and viability of optical links by evaluating the balance between transmitted optical power and received optical power.

To initiate the process of calculating optical power budgets, administrators must first gather relevant information about the optical link, including the characteristics of the fiber optic cable, the optical transmitters, and the receivers involved. CLI commands such as "show-interface-optics" or "display-transmitter-receiver-specifications" can be used to retrieve detailed information about the optical components and parameters of the link.

Once the necessary information is obtained, administrators can begin calculating the optical power budget by determining the optical power levels at various points along the optical path. CLI commands like "measure-optical-power" or "calculate-optical-attenuation" facilitate the measurement of optical power levels at the transmitter output, the receiver input, and any intermediate points in the optical link.

The next step in the calculation process involves assessing the optical losses incurred by the fiber optic cable and other components in the optical path. CLI commands such as "analyze-optical-loss" or "evaluate-fiber-attenuation" assist administrators in quantifying the optical losses due to fiber attenuation, connector losses, splicing losses, and other factors that contribute to signal degradation.

Once the optical losses are determined, administrators can calculate the total optical power budget for the link by subtracting the cumulative losses from the optical power available at the transmitter output. CLI commands like "calculate-optical-power-budget" or "determine-link-margin" help administrators perform these calculations and assess the margin of safety or headroom available in the optical link.

Additionally, administrators may need to consider factors such as safety margins, signal-to-noise ratios, and environmental conditions when calculating optical power budgets to ensure reliable and robust performance of the optical link. CLI commands such as "adjust-optical-power-budget" or "account-for-safety-margins" enable administrators to account for these factors and make adjustments to the optical power budget calculations accordingly.

Furthermore, administrators can use CLI commands to perform simulations or predictive analyses to assess the impact of potential changes or upgrades to the optical network on the optical power budgets. By using commands like "simulate-optical-network-changes" or "predict-optical-power-performance," administrators can evaluate the feasibility and implications of introducing new optical components, increasing link distances, or changing network configurations.

In summary, calculating optical power budgets is a critical aspect of optical networking, enabling administrators to evaluate the performance and reliability of optical links and ensure the successful transmission of data over fiber optic cables. CLI commands provide administrators with the tools and capabilities to gather relevant information, measure optical power levels, assess optical losses, and calculate optical power budgets accurately. By leveraging CLI commands and following best practices in optical power budget calculations, administrators can optimize the performance and efficiency of their optical networks and maintain reliable communication infrastructure.

Analyzing link budgets is a fundamental aspect of designing and maintaining reliable data transmission over communication links. CLI commands play a pivotal role in deploying this technique, providing network administrators with the tools to assess the performance and viability of communication links by evaluating the balance between transmitted power and received power.

To begin the process of analyzing link budgets, administrators first need to gather essential information about the communication link, including the characteristics of the transmitter, the medium (such as fiber optic cable), and the receiver. CLI commands like "show-transmitter-specs," "display-fiber-characteristics," and "view-receiver-specifications" can be utilized to retrieve detailed information about these components and parameters.

Once the relevant information is obtained, administrators can proceed with calculating the link budget by assessing the power levels at different stages of the communication link. CLI commands such as "measure-transmitted-power," "evaluate-received-power," and "calculate-link-margin" enable administrators to determine the transmitted power, received power, and the margin of safety or headroom available in the link.

In addition to measuring power levels, administrators must also account for losses incurred by the communication medium and other components in the link. CLI commands like "analyze-medium-losses," "quantify-connector-losses," and "evaluate-amplification-gains" assist administrators in quantifying these losses and gains accurately.

Furthermore, administrators may need to consider factors such as distance, modulation techniques, signal-to-noise ratio requirements, and environmental conditions when analyzing link budgets to ensure reliable data transmission. CLI commands such as "adjust-link-budget-parameters" or "account-for-noise-factors" enable administrators to incorporate these considerations into their analysis and make adjustments as needed.

Moreover, administrators can use CLI commands to perform simulations or predictive analyses to assess the impact of potential changes or upgrades to the communication link on its performance. Commands like "simulate-link-changes" or "predict-link-performance" facilitate administrators in evaluating the feasibility and implications of introducing new components, increasing link distances, or modifying modulation schemes.

Additionally, administrators can leverage CLI commands to monitor and troubleshoot communication links in real-time, allowing them to detect and address issues promptly to maintain optimal performance. Commands such as "monitor-link-performance" or "troubleshoot-link-issues" provide administrators with insights into the health and stability of the link and enable them to take corrective actions as necessary.

In summary, analyzing link budgets is essential for ensuring reliable data transmission over communication links, and CLI commands are indispensable tools for deploying this technique effectively. By leveraging CLI commands and following best practices in link budget analysis, administrators can optimize the performance and efficiency of their communication networks, leading to improved reliability and overall network performance.

Chapter 8: Environmental Considerations and Protection Methods

In the realm of networking and data transmission, safeguarding infrastructure against physical damage and environmental factors is paramount for maintaining operational continuity and ensuring the longevity of equipment. CLI commands serve as indispensable tools for deploying protective measures and mitigating risks associated with various threats, ranging from natural disasters to human errors.

One of the foremost concerns in protecting network infrastructure is shielding against physical damage, which can arise from diverse sources such as accidental impacts, vandalism, or construction activities. Administrators can employ CLI commands like "configure-access-control" to restrict unauthorized access to critical infrastructure areas, thus minimizing the risk of intentional harm. Additionally, commands such as "enable-intrusion-detection" and "set-up-video-surveillance" enable administrators to monitor premises for suspicious activities and respond swiftly to potential security breaches.

Moreover, safeguarding network components against environmental factors, including temperature fluctuations, humidity, dust, and water ingress, is essential for preventing equipment malfunction and downtime. CLI commands like "configure-environmental-monitoring" allow administrators to deploy sensors and monitoring systems to track environmental conditions in data centers and network facilities. By monitoring parameters such as temperature and humidity levels, administrators can detect anomalies early and take preemptive actions to prevent damage to equipment.

Deploying physical barriers and protective enclosures is another strategy for shielding network infrastructure from environmental hazards. CLI commands like "install-enclosures" and "configure-access-control-lists" facilitate the implementation of access control measures and security protocols to restrict entry to sensitive areas. Additionally, administrators can utilize commands such as "configure-firewall-rules" to define rules for filtering incoming and outgoing traffic, thereby fortifying network defenses against cyber threats and unauthorized access attempts.

Furthermore, proactive maintenance and inspection are essential components of any comprehensive protection strategy, allowing administrators to identify and address potential vulnerabilities before they escalate into critical issues. CLI commands like "schedule-regular-inspections" and "perform-equipment-checks" enable administrators to establish maintenance schedules and conduct routine inspections of network infrastructure, ensuring that equipment remains in optimal condition and capable of withstanding environmental stressors.

In addition to physical safeguards, administrators must also implement robust backup and disaster recovery mechanisms to mitigate the impact of unforeseen events such as natural disasters, power outages, or equipment failures. CLI commands like "configure-backup-and-recovery" facilitate the setup of automated backup routines and replication processes to ensure data integrity and availability in the event of a catastrophe. Moreover, commands such as "test-disaster-recovery-plan" enable administrators to validate the efficacy of their recovery strategies through simulated disaster scenarios.

Furthermore, educating personnel about best practices for handling equipment and responding to emergencies is essential for fostering a culture of safety and resilience within an organization. CLI commands like "schedule-employee-training" facilitate the organization of training sessions and workshops to familiarize staff with safety protocols, emergency procedures, and incident response strategies. By empowering personnel with the knowledge and skills needed to identify and address potential risks, organizations can enhance their overall resilience and mitigate the impact of adverse events on network operations.

In summary, protection against physical damage and environmental factors is a critical aspect of maintaining the reliability and resilience of network infrastructure. CLI commands play a vital role in deploying protective measures, monitoring environmental conditions, and responding effectively to security threats and emergencies. By leveraging CLI commands and adopting a proactive approach to risk management, organizations can safeguard their infrastructure and ensure uninterrupted operation of critical systems in the face of diverse threats and challenges.

In the realm of network infrastructure deployment, ensuring the integrity of connections and protecting equipment from environmental elements are critical for maintaining reliable operations. CLI commands play a pivotal role in implementing sealing and weatherproofing solutions to safeguard equipment and connections against moisture ingress, temperature variations, and other environmental hazards.

One fundamental aspect of sealing and weatherproofing involves protecting outdoor cable connections and terminations from moisture and humidity, which can cause corrosion and signal degradation over time. CLI commands such as "apply-weatherproofing-sealant" facilitate the application of weatherproofing compounds or gels to cable connectors and splices, forming a protective barrier against water intrusion. These compounds are often silicone-based and are applied using specialized tools such as caulking guns or syringes to ensure precise and thorough coverage.

Additionally, CLI commands like "install-weatherproof-enclosures" enable administrators to deploy weatherproof enclosures or junction boxes to house outdoor equipment and cable terminations. These enclosures are designed to withstand exposure to harsh environmental conditions and provide a sealed environment to protect sensitive components from moisture, dust, and debris. They often feature gasketed lids and cable entry ports with compression seals to ensure a watertight seal and prevent water ingress.

Moreover, CLI commands such as "configure-environmental-monitoring" allow administrators to set up monitoring systems to track environmental conditions at outdoor deployment sites. These systems may include sensors for measuring temperature, humidity, and moisture levels, as well as alarms or notifications to alert administrators of any deviations from acceptable parameters. By monitoring environmental conditions in real-time, administrators can proactively address potential issues and take preventive measures to protect equipment from damage.

Furthermore, deploying outdoor-rated cables with robust insulation and jacketing materials is essential for withstanding exposure to UV radiation, extreme temperatures, and mechanical stress. CLI commands like "select-outdoor-rated-cables" facilitate the selection of cables specifically designed for outdoor applications, featuring ruggedized construction and UV-resistant jackets to ensure long-term reliability in harsh environments. Additionally, commands such as "bury-cables-at-appropriate-depth" provide guidance on burying cables underground at the proper depth to protect them from physical damage and environmental hazards.

In addition to outdoor deployments, sealing and weatherproofing solutions are also essential for protecting indoor equipment and cable terminations from environmental factors such as humidity and temperature fluctuations. CLI commands like "install-cable-management-systems" enable administrators to organize and secure cables within equipment racks or cabinets, minimizing the risk of accidental damage or disconnection. Additionally, commands such as "apply-cable-grommets" facilitate the installation of grommets or cable entry seals to prevent air leakage and maintain the integrity of environmental controls within indoor spaces.

Moreover, CLI commands such as "configure-humidity-control-systems" allow administrators to set up humidity control systems within indoor facilities to regulate moisture levels and prevent condensation, which can lead to equipment corrosion and malfunction. These systems may include dehumidifiers, air conditioners, or humidity sensors coupled with automated controls to maintain optimal environmental conditions.

Furthermore, conducting regular inspections and maintenance activities is essential for ensuring the ongoing effectiveness of sealing and weatherproofing solutions. CLI commands like "schedule-regular-inspections" facilitate the establishment of maintenance schedules and reminders for inspecting equipment enclosures, cable connections, and environmental monitoring systems. Additionally, commands such as "perform-visual-inspections" provide guidance on conducting visual inspections to identify signs of wear, damage, or deterioration that may compromise the effectiveness of weatherproofing measures.

In summary, implementing sealing and weatherproofing solutions is crucial for protecting network infrastructure and equipment from environmental hazards. CLI commands enable administrators to deploy a range of protective measures, including weatherproofing compounds, enclosures, outdoor-rated cables, and humidity control systems, to safeguard equipment and connections against moisture ingress, temperature variations, and other environmental threats. By leveraging CLI commands and adopting proactive maintenance practices, organizations can ensure the reliability and longevity of their network infrastructure in diverse operating conditions.

Chapter 9: Compliance and Regulatory Standards in Fiber Optic Infrastructure

In the realm of telecommunications and network infrastructure, adherence to industry standards and regulations is paramount to ensure interoperability, reliability, and compliance with legal requirements. CLI commands play a pivotal role in implementing and verifying adherence to these standards and regulations across various facets of network deployment and operation.

One fundamental aspect of industry standards and regulations pertains to cable specifications and performance requirements. CLI commands such as "check-cable-specifications" enable administrators to verify compliance with standards such as TIA/EIA-568 for structured cabling systems or ANSI/TIA-492 for fiber optic cable specifications. These commands facilitate the validation of cable characteristics such as transmission speed, bandwidth capacity, and signal attenuation, ensuring that deployed cables meet the required performance criteria for their intended applications.

Moreover, CLI commands like "validate-testing-procedures" facilitate the implementation of testing procedures mandated by industry standards and regulations to verify the integrity and performance of network infrastructure. For instance, commands may include "run-cable-certification-tests" to conduct comprehensive testing of copper or fiber optic cables using specialized certification testers. These testers measure parameters such as attenuation, crosstalk, and return loss to assess cable performance and ensure compliance with relevant standards.

Additionally, CLI commands such as "enforce-security-protocols" enable administrators to enforce security protocols mandated by regulatory frameworks such as the Health Insurance Portability and Accountability Act (HIPAA) or the Payment Card Industry Data Security Standard (PCI DSS). These commands facilitate the configuration of encryption, access controls, and audit logging mechanisms to protect sensitive data and ensure compliance with regulatory requirements regarding data privacy and security.

Furthermore, CLI commands play a crucial role in implementing protocols and technologies mandated by industry standards to address specific operational requirements. For example, commands like "configure-quality-of-service" facilitate the implementation of Quality of Service (QoS) mechanisms defined in standards such as IEEE 802.1Q for prioritizing traffic and ensuring predictable performance in converged networks. Similarly, commands such as "deploy-network-segmentation" enable the segmentation of networks into virtual LANs (VLANs) or subnets to enhance security and optimize network traffic management, aligning with best practices outlined in industry standards.

Moreover, CLI commands like "audit-compliance-checks" facilitate the auditing of network configurations and policies to assess compliance with regulatory frameworks such as the General Data Protection Regulation (GDPR) or the Sarbanes-Oxley Act (SOX). These commands enable administrators to conduct automated checks and generate compliance reports to demonstrate adherence to regulatory requirements and address any non-compliance issues promptly.

In addition to network configuration and security, CLI commands are essential for implementing standards-compliant procedures for network management and monitoring. For instance, commands such as "configure-logging-and-alerting" facilitate the configuration of syslog servers and SNMP traps to capture event logs and alert notifications for proactive monitoring and troubleshooting. Similarly, commands like "enforce-auditing-standards" enable the implementation of audit trails and change management procedures to track configuration changes and ensure accountability in accordance with industry standards.

Furthermore, CLI commands play a crucial role in facilitating compliance with environmental regulations and sustainability initiatives related to network infrastructure deployment and operation. For example, commands such as "monitor-power-consumption" enable administrators to track energy usage and optimize power management settings to minimize environmental impact and align with standards such as the Leadership in Energy and Environmental Design (LEED) certification requirements.

In summary, CLI commands are indispensable tools for implementing and ensuring compliance with industry standards and regulations across various aspects of network deployment and operation. From cable specifications and testing procedures to security protocols and regulatory compliance checks, CLI commands enable administrators to enforce standards-compliant practices and maintain the integrity, security, and reliability of network infrastructure in accordance with established industry guidelines and legal requirements.

In the realm of network infrastructure deployment and operation, ensuring compliance with safety and performance standards is paramount to safeguarding personnel, protecting assets, and maintaining operational efficiency. CLI commands serve as indispensable tools for implementing and verifying adherence to these standards across various facets of network management and operation.

One crucial aspect of ensuring compliance with safety standards involves conducting regular inspections and audits of network infrastructure to identify potential hazards and mitigate risks. CLI commands such as "check-hardware-health" enable administrators to monitor the condition of network equipment and detect any signs of physical damage or malfunction that may compromise safety. By routinely running these commands, administrators can proactively identify and address issues to ensure that network infrastructure complies with safety regulations and industry best practices.

Moreover, CLI commands play a crucial role in implementing and enforcing safety protocols and procedures mandated by regulatory frameworks such as the Occupational Safety and Health Administration (OSHA) or the National Electrical Code (NEC). For example, commands like "configure-arc-flash-protection" facilitate the implementation of measures to mitigate the risk of arc flash incidents, such as setting appropriate trip settings on circuit breakers or installing protective barriers around high-voltage equipment. By deploying these commands, administrators can enhance workplace safety and minimize the likelihood of accidents or injuries.

Additionally, CLI commands are essential for enforcing performance standards and ensuring the optimal operation of network infrastructure. Commands such as "monitor-network-performance" enable administrators to assess key performance indicators (KPIs) such as bandwidth utilization, latency, and packet loss to gauge the overall health and performance of the network. By continuously monitoring these metrics, administrators can identify performance bottlenecks or anomalies and take proactive measures to optimize network performance and comply with performance standards. Furthermore, CLI commands facilitate the implementation of Quality of Service (QoS) mechanisms to prioritize traffic and ensure the consistent delivery of critical applications and services. For instance, commands like "configure-qos-policies" enable administrators to define traffic classification criteria and assign appropriate QoS parameters such as priority levels or bandwidth allocations. By applying these commands, administrators can meet performance requirements for mission-critical applications while maintaining compliance with service-level agreements (SLAs) and industry standards.

Moreover, CLI commands are instrumental in verifying compliance with regulatory requirements related to data privacy and security. For example, commands such as "audit-access-controls" enable administrators to assess the effectiveness of access control measures and permissions assigned to users or devices. By conducting regular audits using these commands, administrators can ensure that sensitive data is protected against unauthorized access or disclosure, thereby complying with regulations such as the General Data Protection Regulation (GDPR) or the Health Insurance Portability and Accountability Act (HIPAA).

In addition to safety and performance standards, CLI commands facilitate compliance with environmental regulations and sustainability initiatives aimed at reducing energy consumption and minimizing carbon emissions. Commands such as "monitor-power-consumption" enable administrators to track energy usage and identify opportunities for optimizing power management settings or deploying energy-efficient hardware. By implementing these commands, organizations can reduce their environmental footprint and align with standards such as the Energy Star certification requirements or the Leadership in Energy and Environmental Design (LEED) criteria.

In summary, CLI commands play a crucial role in ensuring compliance with safety and performance standards across various aspects of network infrastructure management and operation. From conducting safety inspections and enforcing protocols to monitoring performance metrics and verifying regulatory compliance, CLI commands empower administrators to maintain a safe, efficient, and compliant network environment in accordance with established industry standards and regulatory requirements.

Chapter 10: Project Management for Fiber Optic Deployment

Planning and scheduling fiber optic projects is a multifaceted process that involves meticulous attention to detail and adherence to established methodologies to ensure successful implementation. CLI commands play a crucial role in this process by facilitating efficient project management, resource allocation, and timeline optimization.

One of the initial steps in planning fiber optic projects is conducting a comprehensive site survey to assess the existing infrastructure, identify potential obstacles, and determine the scope of work required. CLI commands such as "conduct-site-survey" enable project managers to gather essential information about site conditions, including cable pathways, environmental factors, and accessibility constraints. By running commands to collect data on site topology and layout, project teams can make informed decisions regarding cable routing, equipment placement, and installation methodologies.

Once the site survey is complete, project managers can use CLI commands to create detailed project plans and schedules. Commands like "create-project-schedule" allow for the development of timelines that outline key milestones, tasks, and dependencies. By specifying start and end dates for each task and allocating resources accordingly, project managers can effectively coordinate the activities of various teams and ensure that project objectives are met within the designated timeframe.

Furthermore, CLI commands such as "assign-resources" facilitate the allocation of personnel, equipment, and materials to specific tasks within the project schedule. By assigning skilled technicians, specialized tools, and necessary components to each phase of the project, managers can optimize resource utilization and minimize potential delays or bottlenecks. Additionally, commands for resource tracking and management enable real-time monitoring of resource availability and utilization, allowing for timely adjustments to the project plan as needed.

In addition to resource allocation, CLI commands are instrumental in managing communication and collaboration among project stakeholders. Commands like "schedule-meetings" and "send-notifications" enable project managers to schedule regular meetings, disseminate updates, and communicate progress to team members, clients, and other relevant parties. By fostering transparent communication channels and ensuring that all stakeholders are informed of project developments, CLI commands contribute to the overall efficiency and success of fiber optic projects.

Moreover, CLI commands play a vital role in risk management and contingency planning during the project lifecycle. Commands for risk assessment and mitigation enable project managers to identify potential risks, assess their impact on project objectives, and develop strategies to mitigate or address them proactively. By running commands to analyze risk factors such as inclement weather, equipment failures, or resource constraints, project teams can implement preventive measures and contingency plans to minimize disruptions and ensure project continuity.

Furthermore, CLI commands facilitate progress tracking and performance monitoring throughout the project duration. Commands such as "track-progress" and "generate-reports" enable project managers to monitor task completion, track milestone achievements, and evaluate overall project performance against predefined metrics. By generating reports on key performance indicators (KPIs) such as schedule adherence, resource utilization, and budget variance, managers can assess project health and identify areas for improvement or optimization.

Additionally, CLI commands support compliance with regulatory requirements and industry standards governing fiber optic projects. Commands for documentation management and version control enable project teams to maintain accurate records of project documentation, including design specifications, permits, and compliance certificates. By ensuring that project documentation is up-to-date and easily accessible, CLI commands help mitigate risks associated with non-compliance and facilitate regulatory audits or inspections.

In summary, CLI commands are essential tools for planning and scheduling fiber optic projects, enabling project managers to streamline project management processes, optimize resource allocation, and mitigate risks effectively. By leveraging CLI commands for site surveys, project scheduling, resource management, communication, risk assessment, progress tracking, and compliance documentation, project teams can ensure the successful execution of fiber optic projects within budget, scope, and timeline constraints.

Managing resources and stakeholder communication is a critical aspect of project management, particularly in the dynamic and complex environment of fiber optic projects. CLI commands offer valuable capabilities for efficiently allocating resources, coordinating tasks, and facilitating communication among project stakeholders to ensure seamless project execution.

One of the primary challenges in resource management is ensuring optimal utilization of personnel, equipment, and materials throughout the project lifecycle. CLI commands such as "assign-resources" and "allocate-equipment" enable project managers to assign specific tasks to team members and allocate necessary resources based on their skills and availability. By leveraging these commands, managers can ensure that the right resources are allocated to the right tasks at the right time, thereby maximizing efficiency and productivity.

Furthermore, CLI commands play a crucial role in tracking resource utilization and monitoring project progress in real-time. Commands like "track-resources" and "monitor-progress" provide visibility into resource allocation and task completion status, allowing project managers to identify potential bottlenecks or resource shortages early on and take corrective actions as needed. By continuously monitoring resource utilization metrics, managers can proactively address issues and ensure that project milestones are achieved according to schedule.

Effective communication is another key component of successful project management, especially in projects involving multiple stakeholders with diverse roles and responsibilities. CLI commands offer various functionalities for facilitating communication, including scheduling meetings, sending notifications, and sharing project updates. For instance, commands such as "schedule-meeting" and "send-notification" allow project managers to coordinate meetings with stakeholders, disseminate important information, and provide timely updates on project progress.

Moreover, CLI commands enable project managers to customize communication channels and workflows to suit the needs of different stakeholders. Commands for creating communication groups and defining access permissions ensure that relevant information is shared with the appropriate individuals or teams. By segmenting stakeholders based on their roles and responsibilities, managers can streamline communication processes and avoid information overload or miscommunication.

In addition to facilitating communication within the project team, CLI commands also support external communication with clients, vendors, and other external stakeholders. Commands for generating reports, sharing project documentation, and providing status updates enable project managers to keep external stakeholders informed about project developments and milestones. By maintaining transparent communication channels, managers can build trust and confidence among stakeholders and foster productive collaborations.

Furthermore, CLI commands offer functionalities for documenting and archiving communication logs, ensuring accountability and traceability throughout the project lifecycle. Commands for recording meeting minutes, documenting decisions, and tracking action items enable project managers to maintain comprehensive records of project communication and decision-making processes. These records serve as valuable references for future audits, reviews, or project retrospectives.

Additionally, CLI commands support compliance with regulatory requirements and industry standards governing communication and documentation practices. Commands for version control, document management, and access control ensure that project documentation remains accurate, up-to-date, and accessible to authorized personnel. By adhering to established communication protocols and documentation standards, project teams can mitigate risks associated with miscommunication, misunderstandings, or information gaps.

In summary, managing resources and stakeholder communication is essential for the successful execution of fiber optic projects, and CLI commands play a crucial role in facilitating these processes. By leveraging CLI commands for resource allocation, progress tracking, communication facilitation, and documentation management, project managers can streamline project workflows, optimize resource utilization, and ensure effective collaboration among project stakeholders.

BOOK 4
CUTTING-EDGE FIBER OPTICS
EMERGING TECHNOLOGIES AND FUTURE TRENDS IN
NETWORKING

ROB BOTWRIGHT

Chapter 1: Introduction to Emerging Fiber Optic Technologies

Recent advancements and innovations in the field of fiber optics have revolutionized communication networks, data transmission, and various industries reliant on high-speed connectivity. CLI commands such as "update-version" and "install-package" facilitate the deployment of these innovations, ensuring seamless integration into existing infrastructure and systems. One significant advancement is the development of advanced modulation techniques, including coherent detection and quadrature amplitude modulation (QAM), which enable higher data rates and improved spectral efficiency in optical communication systems. These techniques, combined with sophisticated signal processing algorithms, have significantly enhanced the capacity and performance of fiber optic networks, enabling them to meet the ever-increasing demand for bandwidth-intensive applications. Moreover, the emergence of software-defined networking (SDN) and network function virtualization (NFV) has transformed the way network infrastructures are designed, deployed, and managed. CLI commands such as "configure-SDN" and "deploy-VNF" enable network operators to dynamically allocate resources, optimize traffic routing, and deploy virtualized network functions, leading to greater flexibility, scalability, and cost-efficiency. Additionally, advancements in optical fiber manufacturing and materials science have resulted in the development of novel fiber types with enhanced properties, such as reduced signal attenuation, increased bandwidth, and improved resistance to environmental factors. CLI commands such as "manufacture-fiber" and "test-fiber-properties" facilitate the production and testing of these advanced fiber types, ensuring their reliability and performance in real-world applications. Furthermore, the integration of fiber optics with emerging technologies such as artificial intelligence (AI), machine learning (ML), and the Internet of Things (IoT) is opening up new possibilities for smart and autonomous systems. CLI commands such as "integrate-AI" and "deploy-IoT-sensors" enable the seamless integration of fiber optic networks with AI-powered analytics platforms and IoT devices, enabling real-time monitoring, predictive maintenance, and intelligent decision-making. Moreover, recent advancements in fiber optic sensing technologies have expanded the scope of applications beyond telecommunications to include areas such as structural health monitoring, environmental sensing, and medical diagnostics. CLI commands such as "deploy-sensing-system" and "analyze-sensor-data" facilitate the deployment and management of fiber optic sensor networks, enabling accurate and reliable measurement of various physical parameters in diverse environments. Additionally, ongoing research and development efforts are focused on pushing the boundaries of fiber optic technology, with initiatives such as space-based optical communications, quantum communication networks, and ultra-high-capacity transmission systems. CLI commands such as "initiate-research-project" and "collaborate-with-partners" support these efforts by providing tools for project management, collaboration, and resource allocation. Overall, recent advancements and innovations in fiber optics are driving unprecedented progress in telecommunications, data networking, and sensing applications, offering new opportunities for connectivity, efficiency, and intelligence in the digital age.

Emerging technologies play a pivotal role in modern communication, facilitating the evolution of networks and the seamless exchange of information across various platforms and devices. CLI commands such as "deploy-technology" and "configure-network" are instrumental in implementing these advancements, ensuring their integration into existing infrastructures. One such technology is 5G, the fifth generation of mobile networks, which promises ultra-fast data speeds, low latency, and massive connectivity, revolutionizing mobile communication and enabling new applications such as autonomous vehicles, augmented reality, and remote surgery. CLI commands like "upgrade-to-5G" and "configure-latency-settings" enable network operators to deploy and optimize 5G networks for enhanced performance and reliability. Additionally, the Internet of Things (IoT) is transforming communication by connecting billions of devices and sensors, enabling data collection, analysis, and automation across various industries. CLI commands such as "connect-IoT-devices" and "deploy-sensor-networks" facilitate the deployment and management of IoT ecosystems, allowing organizations to monitor assets, optimize processes, and deliver personalized services. Moreover, artificial intelligence (AI) and machine learning (ML) are revolutionizing communication by enabling intelligent decision-making, predictive analytics, and natural language processing. CLI commands such as "integrate-AI-algorithms" and "train-ML-models" empower organizations to leverage AI-powered applications for speech recognition, chatbots, and personalized content recommendation, enhancing the user experience and improving operational efficiency. Furthermore, blockchain technology is revolutionizing communication by providing a secure and decentralized framework for data exchange, ensuring transparency, immutability, and trust in transactions. CLI commands such as "implement-blockchain-network" and "validate-transactions" enable organizations to deploy blockchain-based solutions for secure messaging, digital identity verification, and supply chain management. Additionally, edge computing is reshaping communication by bringing computational power closer to the data source, reducing latency and enabling real-time processing for mission-critical applications. CLI commands such as "deploy-edge-computing-nodes" and "optimize-data-processing" empower organizations to build edge computing infrastructure for applications such as video streaming, autonomous vehicles, and industrial automation. Moreover, quantum communication is poised to revolutionize communication security by leveraging the principles of quantum mechanics to enable unbreakable encryption and secure data transmission. CLI commands such as "establish-quantum-communication-link" and "generate-quantum-keys" enable organizations to deploy quantum communication networks for secure communication and data exchange. Additionally, advancements in satellite communication technology are expanding connectivity to remote and underserved areas, bridging the digital divide and enabling access to information and services. CLI commands such as "launch-satellite" and "configure-satellite-network" facilitate the deployment and management of satellite communication systems, providing reliable and ubiquitous connectivity for users worldwide. Furthermore, advances in optical communication technology are driving the development of high-speed and high-capacity fiber optic networks, enabling the transmission of vast amounts of data over long distances with minimal signal degradation. CLI commands such as "upgrade-to-optical-fiber" and "optimize-data-transmission" support the deployment and optimization of fiber optic networks for applications such as broadband internet, cloud computing, and video streaming. Overall, emerging technologies are essential for driving innovation and advancing communication capabilities in the digital age, empowering organizations to deliver faster, more reliable, and more secure communication services to users worldwide.

Chapter 2: Silicon Photonics: Integration and Miniaturization

Silicon photonics is a transformative technology that integrates optical components such as waveguides, modulators, and photodetectors directly onto silicon substrates, enabling the fabrication of highly integrated and compact photonic circuits. The adoption of silicon photonics is driven by its potential to revolutionize various fields, including telecommunications, data centers, and biophotonics. CLI commands such as "design-silicon-photonics" and "simulate-optical-circuits" facilitate the design and optimization of silicon photonic devices, enabling researchers and engineers to harness their full potential. One of the fundamental principles of silicon photonics is the use of silicon as the platform material due to its compatibility with complementary metal-oxide-semiconductor (CMOS) technology, which enables seamless integration with existing electronic circuits. CLI commands such as "fabricate-Si-wafer" and "integrate-CMOS" enable the fabrication and integration of silicon photonic devices with electronic components on a single chip, enabling multifunctional and energy-efficient systems. Another key principle of silicon photonics is the use of optical waveguides to confine and guide light within the silicon substrate. Waveguides can be fabricated using various techniques, including lithography and etching, with CLI commands such as "etch-waveguide" and "inspect-waveguide-quality" facilitating the fabrication and quality control processes. By guiding light along the silicon substrate, waveguides enable the routing and manipulation of optical signals within photonic circuits, forming the basis for various photonic devices and functionalities. Additionally, silicon photonics leverages the electro-optic properties of silicon to modulate and detect light using electrostatic or piezoelectric effects. CLI commands such as "apply-voltage-modulator" and "measure-detector-response" enable researchers to control and characterize the performance of silicon photonic modulators and detectors, essential components for signal generation and detection in photonic systems. Moreover, silicon photonics enables the integration of passive and active optical components on the same chip, including filters, splitters, amplifiers, and lasers, allowing for the realization of complex photonic circuits with enhanced functionality and performance. CLI commands such as "integrate-passive-components" and "test-active-devices" support the integration and testing of these components, ensuring their compatibility and reliability in silicon photonic systems. Furthermore, silicon photonics offers significant advantages in terms of scalability, cost-effectiveness, and compatibility with existing manufacturing processes, making it an attractive platform for mass production and deployment in various applications. CLI commands such as "scale-up-production" and "optimize-manufacturing-process" facilitate the scaling and optimization of silicon photonic fabrication processes, enabling cost-effective and scalable production of photonic devices. Overall, the principles of silicon photonics represent a fundamental paradigm shift in the field of photonics, offering unprecedented opportunities for the development of compact, high-performance, and energy-efficient photonic systems for a wide range of applications.

Advances in integration and miniaturization techniques have played a pivotal role in revolutionizing various fields, from electronics to photonics, enabling the development of compact, multifunctional, and energy-efficient devices and systems. These advances have been driven by the relentless pursuit of higher performance, smaller form factors, and lower power consumption, facilitated by innovations in materials science, fabrication processes, and design methodologies. CLI commands such as "optimize-integration-process" and "minimize-footprint" are instrumental in achieving these goals, allowing engineers and researchers to leverage advanced integration and miniaturization techniques effectively.

One of the key advancements in integration and miniaturization is the development of system-on-chip

(SoC) and system-in-package (SiP) technologies, which enable the integration of multiple functionalities and components onto a single chip or package. This integration reduces the footprint, power consumption, and cost of electronic and photonic systems while improving their performance and reliability. CLI commands such as "design-SoC" and "assemble-SiP" facilitate the design and fabrication of integrated systems, enabling the seamless integration of diverse functionalities such as processing, memory, communication, and sensing.

Another significant advancement is the emergence of 3D integration techniques, which enable the stacking of multiple layers of active and passive components within a single package. This vertical integration increases packing density and interconnectivity, allowing for the realization of highly integrated and compact systems with enhanced performance and functionality. CLI commands such as "implement-3D-integration" and "optimize-interconnect-design" support the deployment and optimization of 3D integration techniques, enabling the fabrication of advanced electronic and photonic devices.

Furthermore, advancements in nanofabrication techniques have enabled the precise control and manipulation of materials and structures at the nanoscale, paving the way for the development of nanoelectronic and nanophotonic devices with unprecedented performance and functionality. CLI commands such as "fabricate-nanodevices" and "characterize-nanomaterials" facilitate the fabrication and characterization of nanoscale components, enabling researchers to explore new phenomena and applications in electronics and photonics.

In addition to electronics, integration, and miniaturization techniques have also revolutionized the field of photonics, enabling the development of compact and efficient photonic devices and systems for various applications. One notable example is silicon photonics, which leverages advanced semiconductor fabrication processes to integrate optical components such as waveguides, modulators, and detectors on a silicon substrate. CLI commands such as "design-silicon-photonics" and "fabricate-photonic-components" support the design and fabrication of silicon photonic devices, enabling the realization of high-performance optical interconnects, sensors, and signal processing systems.

Moreover, advances in microelectromechanical systems (MEMS) technology have enabled the integration of mechanical and optical components on the same chip, enabling the development of microscale actuators, sensors, and switches with a wide range of applications in telecommunications, healthcare, and consumer electronics. CLI commands such as "integrate-MEMS-devices" and "test-microactuators" facilitate the integration and testing of MEMS-based components, enabling the deployment of advanced microsystems for diverse applications.

Overall, advances in integration and miniaturization techniques have transformed the landscape of electronics and photonics, enabling the development of compact, multifunctional, and energy-efficient devices and systems with unprecedented performance and functionality. CLI commands play a crucial role in supporting these advancements, enabling researchers and engineers to design, fabricate, and deploy integrated systems effectively.

Chapter 3: Quantum Key Distribution for Secure Communication

Quantum Key Distribution (QKD) is a revolutionary cryptographic technique that leverages the principles of quantum mechanics to secure communication channels against eavesdropping and interception. Unlike classical encryption methods, which rely on the complexity of mathematical algorithms, QKD exploits the inherent properties of quantum physics, such as the uncertainty principle and the no-cloning theorem, to enable the secure exchange of cryptographic keys between two parties. CLI commands such as "initialize-QKD-system" and "establish-quantum-link" facilitate the deployment and configuration of QKD systems, allowing users to establish secure communication channels over optical fibers or free-space links. One of the fundamental principles of QKD is the use of quantum states, such as single photons or entangled photon pairs, to encode and transmit cryptographic information between the sender and the receiver. CLI commands such as "generate-single-photon" and "measure-photon-state" enable researchers to prepare and characterize the quantum states used in QKD protocols, ensuring their integrity and security. Another key principle of QKD is the concept of quantum entanglement, which allows two or more particles to become intrinsically correlated, regardless of the distance between them. CLI commands such as "create-entangled-pair" and "distribute-entangled-photons" enable the generation and distribution of entangled photon pairs, which can be used to establish secure quantum communication links between distant parties. Additionally, QKD protocols typically rely on the principles of quantum measurement and uncertainty to ensure the security of the exchanged cryptographic keys. CLI commands such as "perform-basis-measurement" and "calculate-quantum-uncertainty" enable users to implement and analyze the quantum measurements required for QKD protocols, verifying the authenticity and confidentiality of the exchanged keys. Furthermore, QKD systems often incorporate advanced quantum technologies, such as quantum memories and quantum repeaters, to extend the range and improve the performance of secure communication networks. CLI commands such as "deploy-quantum-memory" and "configure-quantum-repeater" facilitate the integration and optimization of these quantum technologies within QKD systems, enabling the establishment of secure communication links over long distances. Moreover, QKD protocols are designed to be provably secure against various types of attacks, including intercept-resend attacks and man-in-the-middle attacks, by exploiting the fundamental laws of quantum mechanics. CLI commands such as "simulate-eavesdropping-attack" and "analyze-quantum-security" allow researchers to evaluate the security of QKD protocols under different threat scenarios, ensuring their resilience against potential adversaries. Additionally, QKD has practical applications in various fields, including secure communication networks, financial transactions, and government agencies, where the protection of sensitive information is paramount. CLI commands such as "deploy-QKD-network" and "integrate-QKD-with-secure-systems" enable organizations to implement QKD solutions to safeguard their data and communications against unauthorized access and surveillance. Overall, the fundamentals of Quantum Key Distribution represent a significant breakthrough in the field of cryptography, offering unparalleled security guarantees based on the principles of quantum mechanics.

Applications of Quantum Key Distribution (QKD) in secure communication networks are extensive, offering unprecedented levels of security for sensitive data transmission. CLI commands such as "deploy-QKD-network" and "configure-QKD-protocols" are instrumental in implementing QKD solutions within communication infrastructures. QKD provides a robust framework for establishing secure communication channels by leveraging the principles of quantum mechanics to encrypt and decrypt data. Organizations can utilize QKD to safeguard their confidential information, such as financial transactions, government communications, and classified data, against interception and eavesdropping. CLI commands like "initialize-QKD-system" and "establish-quantum-link" enable the setup and maintenance of QKD connections between network nodes, ensuring the confidentiality and integrity of transmitted data. Moreover, QKD offers unique advantages over traditional cryptographic methods, such as symmetric key encryption and public-key cryptography, by providing unconditional security based on the laws of quantum physics. CLI commands such as "generate-quantum-keys" and "exchange-quantum-signatures" facilitate the generation and exchange of cryptographic keys using QKD protocols, ensuring that only authorized parties can access the encrypted data. Additionally, QKD can be integrated with existing communication protocols and technologies, such as Internet Protocol Security (IPsec) and Virtual Private Networks (VPNs), to enhance the security of data transmission over public and private networks. CLI commands like "integrate-QKD-with-IPsec" and "configure-QKD-based-VPNs" streamline the integration of QKD into existing network infrastructures, allowing organizations to leverage its security benefits without disrupting their operations. Furthermore, QKD enables secure communication in emerging technologies, such as the Internet of Things (IoT) and cloud computing, where data privacy and confidentiality are critical concerns. CLI commands such as "deploy-QKD-for-IoT-devices" and "implement-QKD-in-cloud-networks" support the integration of QKD solutions into IoT devices and cloud-based services, protecting sensitive data from unauthorized access and cyber threats. Moreover, QKD can enhance the security of wireless communication networks, including satellite communication systems and mobile networks, by providing secure key distribution mechanisms that are immune to interception and hacking attempts. CLI commands like "secure-wireless-networks-with-QKD" and "implement-QKD-in-satellite-communications" facilitate the deployment of QKD solutions in wireless communication infrastructures, ensuring the confidentiality and integrity of transmitted data. Additionally, QKD has applications in critical infrastructure protection, such as energy grids, transportation networks, and healthcare systems, where secure communication is essential for ensuring the reliability and safety of operations. CLI commands such as "integrate-QKD-into-critical-infrastructure" and "protect-energy-grids-with-QKD" support the implementation of QKD solutions in critical infrastructure networks, mitigating the risks associated with cyberattacks and data breaches. Furthermore, QKD can play a vital role in securing communication links for remote sensing and surveillance applications, such as defense and national security, where real-time data transmission and secure communication are paramount. CLI commands like "deploy-QKD-for-remote-surveillance" and "secure-defense-networks-with-QKD" enable the integration of QKD into remote sensing and surveillance systems, ensuring the confidentiality and integrity of sensitive information. Overall, the applications of QKD in secure communication networks are diverse and far-reaching, offering robust solutions for protecting data privacy and confidentiality in various domains and industries.

Chapter 4: Nonlinear Optics and All-Optical Signal Processing

Nonlinear optical phenomena and effects play a crucial role in various applications across science and technology, leveraging the unique properties of optical materials to enable advanced functionalities and capabilities. These phenomena arise when the response of a material to an optical field is nonlinear, meaning it does not follow a simple proportional relationship with the applied field. Instead, the material's response becomes dependent on the intensity of the optical field, leading to a range of fascinating effects and applications. CLI commands such as "generate-optical-field" and "measure-nonlinear-response" are instrumental in studying and exploiting these phenomena in practical systems, allowing researchers and engineers to manipulate light in novel ways for diverse applications.

One of the fundamental nonlinear optical phenomena is optical harmonic generation, where photons of different frequencies are generated when intense light interacts with a nonlinear medium. This process, described by CLI commands such as "apply-nonlinear-medium" and "observe-harmonic-generation," enables the generation of coherent light at frequencies that are integer multiples of the incident frequency, facilitating applications such as frequency conversion, laser spectroscopy, and microscopy.

Another significant nonlinear effect is the Kerr effect, where the refractive index of a material changes in response to the intensity of an optical field. CLI commands such as "apply-Kerr-medium" and "measure-refractive-index-change" allow researchers to manipulate and characterize this effect, enabling applications such as all-optical switching, optical signal processing, and nonlinear microscopy. Additionally, the Kerr effect is exploited in Kerr-lens mode-locked lasers, CLI command "activate-Kerr-lens-locking," which generate ultrashort optical pulses for applications in ultrafast spectroscopy, telecommunications, and laser micromachining.

Nonlinear optical phenomena also include optical solitons, which are self-sustaining wave packets that propagate without changing shape due to a balance between nonlinear and dispersive effects. CLI commands such as "create-optical-soliton" and "study-soliton-propagation" are used to generate and analyze these unique waveforms, which find applications in optical communications, fiber lasers, and mode-locked lasers.

Furthermore, nonlinear optics enables the phenomenon of four-wave mixing (FWM), where photons interact within a nonlinear medium to generate new frequencies through a process akin to parametric amplification. CLI commands such as "initiate-four-wave-mixing" and "analyze-FWM-spectrum" are employed to investigate and harness this effect for applications such as wavelength conversion, optical amplification, and optical signal processing in photonic integrated circuits.

Additionally, nonlinear optical effects play a crucial role in the development of nonlinear optical microscopy techniques, CLI command "deploy-nonlinear-microscope," which offer high-resolution imaging capabilities with contrast mechanisms based on nonlinear interactions between light and matter. These techniques, including two-photon microscopy and harmonic generation microscopy, enable label-free imaging of biological samples, neuroimaging, and studies of cellular dynamics.

Moreover, nonlinear optics finds applications in quantum information processing, CLI command "apply-nonlinear-optics-in-QIP," where nonlinear optical processes are used to manipulate and encode quantum information in photonic qubits. These applications include quantum gates, quantum memories, and quantum key distribution protocols, leveraging the unique properties of nonlinear optical systems for secure and efficient quantum communication.

In summary, nonlinear optical phenomena and effects offer a rich toolbox for manipulating light and enabling a wide range of applications across various fields, including photonics, telecommunications, microscopy, and quantum information processing. Understanding and harnessing these phenomena through CLI commands facilitate the development of advanced optical technologies with enhanced functionality and performance.

All-optical signal processing techniques and applications represent a cutting-edge field within optical communications and photonics, offering innovative ways to manipulate and process optical signals entirely using light itself, without the need for conversion to electronic signals. These techniques leverage the unique properties of nonlinear optical materials and devices to perform various signal processing functions, CLI commands such as "activate-all-optical-processing" and "analyze-signal-processing-results" are essential for deploying and assessing these techniques, allowing researchers and engineers to design advanced optical systems with enhanced performance and functionality.

One of the primary applications of all-optical signal processing is in optical communication networks, where these techniques enable high-speed signal modulation, switching, and routing without the need for costly and power-consuming electronic components. CLI command "deploy-all-optical-switch" allows for the implementation of all-optical switches, which use nonlinear optical effects such as the Kerr effect to route optical signals without conversion to electronic signals, reducing latency and improving network efficiency.

Additionally, all-optical signal processing techniques find applications in optical regeneration, CLI command "enable-optical-regeneration," where they are used to mitigate signal degradation and noise accumulation in long-haul fiber-optic transmission systems. By leveraging nonlinear optical effects such as optical phase conjugation and four-wave mixing, these techniques enable the restoration of optical signals to their original quality, extending the reach and capacity of optical communication networks.

Moreover, all-optical signal processing plays a crucial role in optical data encryption and security, CLI command "apply-all-optical-encryption," where it enables the implementation of secure communication protocols based on optical encryption and decryption techniques. By utilizing nonlinear optical devices such as optical parametric oscillators and nonlinear frequency converters, these techniques provide robust encryption of optical signals, protecting sensitive information from eavesdropping and interception.

Furthermore, all-optical signal processing techniques enable advanced modulation formats and multiplexing schemes in optical communication systems, CLI command "implement-all-optical-multiplexing," allowing for the transmission of multiple data channels over a single optical fiber. Techniques such as optical phase modulation, spectral phase encoding, and optical orthogonal frequency-division multiplexing (OFDM) enable high-capacity and spectrally efficient transmission of data in next-generation optical networks.

In addition to telecommunications, all-optical signal processing techniques find applications in optical sensing and metrology, CLI command "deploy-all-optical-sensing," where they enable high-resolution and sensitive measurements of physical parameters such as temperature, pressure, and strain using optical signals. By exploiting nonlinear effects such as stimulated Brillouin scattering and stimulated Raman scattering, these techniques provide precise and accurate sensing capabilities for various industrial and environmental monitoring applications.

Moreover, all-optical signal processing techniques are essential for implementing complex optical computing and signal processing functions, CLI command "enable-all-optical-computing," where they enable the realization of optical logic gates, Boolean operations, and signal transformations entirely using light. These techniques hold promise for developing ultrafast and energy-efficient optical computing platforms for applications such as machine learning, artificial intelligence, and big data analytics.

Furthermore, all-optical signal processing techniques enable novel approaches to nonlinear microscopy and imaging, CLI command "apply-all-optical-imaging," where they provide enhanced contrast and resolution for imaging biological samples and materials. Techniques such as multiphoton microscopy, coherent anti-Stokes Raman scattering (CARS), and second-harmonic generation (SHG) microscopy enable label-free and high-resolution imaging of biological tissues, cells, and nanoparticles.

Additionally, all-optical signal processing techniques are being explored for applications in quantum information processing, CLI command "explore-all-optical-QIP," where they enable the manipulation and control of quantum states of light for quantum computing, cryptography, and communication. Techniques such as optical quantum gates, quantum teleportation, and quantum key distribution leverage nonlinear optical effects to implement essential building blocks for quantum information processing systems.

In summary, all-optical signal processing techniques and applications represent a rapidly evolving field within photonics and optical communications, offering unprecedented capabilities for manipulating and processing optical signals entirely using light. By leveraging nonlinear optical effects and devices, these techniques enable advanced functionalities and applications across various domains, including telecommunications, sensing, computing, and quantum information processing. Deploying these techniques using CLI commands facilitates the development of next-generation optical systems with enhanced performance, efficiency, and functionality.

Chapter 5: Photonic Crystal Fibers: Novel Structures for Enhanced Performance

Designing and fabricating photonic crystal fibers (PCFs) involves intricate processes aimed at achieving specific optical properties and functionalities, CLI commands such as "design-PCF-structure" and "fabricate-PCF-sample" are crucial for implementing these techniques, allowing researchers and engineers to create custom-designed optical fibers with tailored characteristics for various applications. PCFs are characterized by a unique microstructured design that manipulates the propagation of light through periodic variations in refractive index, enabling novel functionalities such as dispersion engineering, nonlinear optics, and enhanced light-matter interactions.

The design process begins with defining the structural parameters of the PCF, CLI command "define-PCF-geometry," including the diameter and arrangement of air holes or high-index inclusions within the fiber core and cladding regions. Advanced numerical simulations and optimization techniques, CLI command "simulate-PCF-design," are employed to model the light propagation characteristics and optical properties of the PCF structure, allowing researchers to refine the design parameters for specific applications.

Once the desired PCF structure is determined, CLI commands such as "optimize-PCF-design" are used to fine-tune the structural parameters to achieve desired optical properties such as dispersion, nonlinear coefficient, and modal characteristics. Optimization algorithms and computational tools play a crucial role in iteratively adjusting the PCF geometry to meet the performance requirements of the intended application, CLI command "deploy-optimized-PCF-design."

After the design phase is complete, the fabrication process begins, CLI command "initiate-PCF-fabrication," typically using specialized techniques such as stack-and-draw or drilling and fusion splicing. In stack-and-draw fabrication, CLI commands such as "prepare-preform-stack" and "draw-PCF-fiber" are used to assemble a preform stack consisting of multiple glass rods or capillaries arranged according to the desired PCF structure, which is then heated and drawn into a thin fiber.

Alternatively, in drilling and fusion splicing fabrication, CLI commands such as "drill-PCF-holes" and "splice-PCF-components" are employed to create air holes or high-index inclusions in a solid glass rod or preform, followed by fusion splicing with a capillary tube or solid glass rod to form the PCF structure. Each fabrication method offers advantages and challenges in terms of scalability, precision, and control over the PCF geometry and properties.

During the fabrication process, CLI commands such as "monitor-fabrication-process" are used to ensure the quality and integrity of the PCF structure, allowing for real-time monitoring of key parameters such as temperature, pressure, and gas flow rate. Quality control techniques such as optical microscopy, scanning electron microscopy (SEM), and spectroscopic analysis are employed to inspect the fabricated PCF samples for defects, CLI command "inspect-PCF-quality."

Once the PCF fabrication is complete, CLI commands such as "characterize-PCF-properties" are used to evaluate the optical performance and properties of the fabricated PCF samples. Various experimental techniques such as optical spectroscopy, nonlinear optical measurements, and mode analysis are employed to quantify parameters such as dispersion, nonlinear coefficient, modal properties, and light-matter interactions.

The fabricated PCF samples are then integrated into optical systems and devices for various applications, CLI commands such as "integrate-PCF-into-optical-setup."

In telecommunications, CLI command "deploy-PCF-in-data-transmission" PCFs are used for dispersion management, wavelength conversion, and nonlinear signal processing in high-speed fiber-optic communication networks. In sensing and imaging applications, CLI command "apply-PCF-in-sensing" PCFs enable enhanced light-matter interactions, leading to improved sensitivity and resolution in optical sensing and imaging systems.

Furthermore, PCFs find applications in laser technology, CLI command "incorporate-PCF-in-laser-design," where they enable compact and efficient laser sources with tunable wavelengths, ultrashort pulse generation, and high-power operation. PCF-based lasers are utilized in various fields such as medical imaging, material processing, and scientific research.

In summary, designing and fabricating photonic crystal fibers (PCFs) involve a combination of advanced numerical simulations, optimization techniques, and specialized fabrication methods. CLI commands play a crucial role in guiding the design, fabrication, characterization, and integration of PCFs into optical systems for a wide range of applications in telecommunications, sensing, imaging, and laser technology. By leveraging the unique properties of PCFs, researchers and engineers can develop innovative optical devices and systems with enhanced performance and functionality.

Photonic Crystal Fibers (PCFs) have emerged as a revolutionary technology in the field of optical communication, offering unique advantages and enabling a wide range of applications. The design and fabrication of PCFs involve sophisticated techniques aimed at controlling light propagation within the fiber structure to achieve desired optical properties. One such technique is the stack-and-draw method, where a preform consisting of a stack of hollow capillaries arranged in a periodic lattice pattern is drawn into a fiber, CLI command "draw-PCF-preform." Another approach involves using a modified chemical vapor deposition (MCVD) process to deposit high-quality silica layers around air holes in a solid preform, CLI command "fabricate-PCF-MCVD." These fabrication methods allow precise control over the fiber's structural parameters, such as hole size, pitch, and geometry, enabling tailored optical properties suited for specific applications.

The unique structure of PCFs offers several advantages over conventional optical fibers, making them highly desirable for various optical communication applications. One significant advantage is their large mode area, CLI command "analyze-PCF-mode-area," which enables efficient coupling of high-power laser sources and mitigates nonlinear effects such as optical Kerr effect and stimulated Raman scattering, CLI command "mitigate-nonlinear-effects." This property is particularly advantageous in high-power fiber laser systems and optical amplifiers, where PCFs can improve beam quality, increase threshold power, and enhance overall system performance.

Furthermore, PCFs exhibit low dispersion and loss characteristics, CLI command "measure-PCF-dispersion," making them suitable for long-haul telecommunications applications. Their unique dispersion properties can be tailored to achieve specific dispersion profiles, including dispersion-flattened, dispersion-shifted, and ultra-low dispersion designs, CLI command "optimize-PCF-dispersion." These characteristics enable PCFs to support high-speed data transmission over long distances while minimizing signal distortion and dispersion-induced pulse broadening, CLI command "analyze-PCF-loss."

Another advantage of PCFs is their highly customizable modal properties, CLI command "customize-PCF-modes," allowing for tailored modal dispersion and polarization properties. This flexibility enables the development of specialty fibers optimized for polarization-maintaining, single-mode, or multi-mode operation, CLI command "design-specialty-PCF." Additionally, PCFs can support multiple guided modes simultaneously, enabling novel functionalities such as mode-division multiplexing (MDM), CLI command "implement-MDM-PCF," for increasing the data transmission capacity of optical communication systems.

The unique nonlinear properties of PCFs also open up new possibilities for nonlinear optical signal processing and wavelength conversion applications. CLI command "deploy-PCF-nonlinear-optics" enables the implementation of various nonlinear processes, including four-wave mixing (FWM), self-phase modulation (SPM), and cross-phase modulation (XPM), CLI command "enable-nonlinear-effects." These nonlinear effects can be harnessed for applications such as wavelength conversion, signal regeneration, and generation of optical frequency combs, CLI command "generate-optical-frequency-comb."

Moreover, PCFs find applications in emerging fields such as optical sensing and imaging, CLI command "utilize-PCF-sensing-imaging," where their unique optical properties enable high-resolution and sensitive detection of physical parameters such as temperature, pressure, and refractive index. PCFs can be tailored for specific sensing applications by adjusting their structural parameters and modal properties, CLI command "optimize-PCF-sensing-properties," allowing for the development of compact and highly sensitive fiber-optic sensors.

Additionally, PCFs have been explored for applications in quantum communication and quantum information processing, CLI command "explore-PCF-quantum-applications," where their unique properties enable the manipulation and control of quantum states of light. PCFs can support the generation, transmission, and manipulation of single photons and entangled photon pairs, CLI command "generate-entangled-photons," facilitating the development of quantum key distribution (QKD) systems and quantum networks.

In summary, Photonic Crystal Fibers (PCFs) offer a myriad of advantages and applications in optical communication, spanning from telecommunications and fiber lasers to nonlinear optics, sensing, and quantum information processing. Their unique structural and optical properties, CLI command "analyze-PCF-properties," enable tailored solutions for diverse technological challenges, making them indispensable tools in modern optical communication systems. Deploying PCFs in various applications requires careful consideration of their design parameters and fabrication techniques, CLI command "deploy-PCF-application," ensuring optimal performance and functionality for specific use cases.

Chapter 6: Metamaterials in Fiber Optic Communications

Metamaterials represent a groundbreaking advancement in the realm of material science, enabling the creation of substances with properties that transcend those found in nature. These materials are engineered to exhibit unique electromagnetic characteristics not typically observed in natural materials. Metamaterials derive their extraordinary properties from their intricate internal structures, which are engineered at a scale smaller than the wavelength of the electromagnetic waves they interact with. This deliberate manipulation of electromagnetic waves offers unprecedented control over light propagation, making metamaterials highly desirable for various applications in optics, photonics, and beyond.

The concept of metamaterials originated from the desire to manipulate electromagnetic waves at subwavelength scales. Traditional materials, governed by the laws of classical physics, have inherent limitations in controlling electromagnetic waves with wavelengths larger than the material's dimensions. Metamaterials, on the other hand, exploit the interactions between subwavelength structures and incident electromagnetic waves to achieve unique optical properties. This capability has opened up a vast array of possibilities for designing novel devices with functionalities unattainable using conventional materials.

One notable aspect of metamaterials is their ability to exhibit negative refractive index, a property not found in natural materials. In ordinary materials, such as glass or water, light bends toward the normal when passing from one medium to another, in accordance with Snell's law. However, metamaterials can exhibit negative refraction, causing light to bend in the opposite direction. This phenomenon arises from the engineered subwavelength structures within the metamaterial, which manipulate the phase and group velocities of light in unconventional ways.

The development of metamaterials has led to the realization of several groundbreaking applications in various fields. In optics and photonics, metamaterials have enabled the creation of superlenses capable of imaging objects at resolutions beyond the diffraction limit of conventional lenses. By exploiting negative refraction, these superlenses can focus light to spot sizes much smaller than the wavelength of light used for imaging, opening up new possibilities for high-resolution microscopy and nanoscale imaging.

Moreover, metamaterials have revolutionized the field of antenna design by enabling the creation of ultrathin, conformal antennas with unprecedented properties. Traditional antennas are limited in their performance by their size and shape, which are dictated by the wavelength of the electromagnetic waves they emit or receive. Metamaterial-based antennas, however, can achieve functionalities such as beam steering, polarization control, and broadband operation in compact form factors, making them ideal for applications in wireless communication systems, radar, and sensing.

In addition to optics and antenna technology, metamaterials have found applications in various other domains, including cloaking devices, energy harvesting, and acoustic engineering. Metamaterial-based cloaking devices can render objects invisible to certain wavelengths of light or sound by bending incident waves around them, effectively concealing their presence from detection. This capability has implications for military stealth technology, as well as for creating more efficient acoustic barriers and vibration damping systems in architectural and automotive applications.

Deploying metamaterials in practical devices often involves sophisticated fabrication techniques and precise control over material properties at nanometer scales. Advanced lithographic methods, such as electron beam lithography and focused ion beam milling, are commonly used to pattern metamaterial structures with subwavelength precision on substrates. Additionally, techniques such as atomic layer deposition and chemical vapor deposition are employed to deposit thin films of metamaterial constituents, allowing for the precise control of material composition and properties.

In summary, metamaterials represent a paradigm shift in material science, offering unprecedented control over light and electromagnetic waves. Their unique properties and functionalities have enabled transformative advances across a wide range of fields, from optics and photonics to telecommunications and beyond. As research in metamaterials continues to progress, further innovations and applications are expected to emerge, driving forward technological advancements and shaping the future of science and engineering.

Metamaterials have emerged as a promising technology for enhancing various aspects of fiber optic communication systems, offering unique capabilities that traditional materials cannot match. These engineered materials, with their tailored electromagnetic properties, hold the potential to revolutionize the way optical signals are transmitted, manipulated, and detected within fiber optic networks. By integrating metamaterials into fiber optic components and devices, researchers and engineers aim to overcome existing limitations and unlock new functionalities, paving the way for more efficient, reliable, and versatile communication systems.

One area where metamaterials show significant promise is in the development of novel fiber optic components with enhanced performance characteristics. Traditional optical fibers, composed of silica glass, are limited by their intrinsic properties, such as dispersion and nonlinearity, which can degrade signal quality and limit transmission distances. Metamaterial-based fibers offer the possibility of tailoring these properties to suit specific application requirements, thereby mitigating performance limitations and enabling the realization of high-performance optical communication systems.

For instance, metamaterials can be used to engineer fibers with tailored dispersion profiles, allowing for the precise control of signal propagation characteristics. By manipulating the refractive index profile of the fiber using metamaterial structures, dispersion can be minimized or even reversed, enabling the design of dispersion-compensating fibers that mitigate signal distortion over long transmission distances. This capability is essential for high-speed data transmission applications, where maintaining signal integrity is paramount.

Moreover, metamaterials can be incorporated into fiber optic amplifiers to enhance their performance and efficiency. Erbium-doped fiber amplifiers (EDFAs), widely used in optical communication systems to boost signal power, can benefit from metamaterial-based designs that improve gain, noise figure, and bandwidth. By tailoring the local electromagnetic environment around the erbium ions using metamaterial structures, it is possible to optimize amplifier performance and achieve higher signal-to-noise ratios, enabling longer transmission distances and higher data rates.

In addition to amplification, metamaterials offer opportunities for enhancing the functionality of fiber optic sensors used in communication networks. Fiber optic sensors are crucial for monitoring environmental conditions, detecting intrusions, and ensuring the integrity of the network infrastructure. By integrating metamaterials into the sensing elements of fiber optic sensors, it is possible to enhance sensitivity, resolution, and selectivity, enabling the detection of smaller changes in the measured parameters with higher precision.

Furthermore, metamaterials can be utilized to overcome limitations related to signal routing and manipulation in fiber optic networks. Traditional optical switches and routers rely on bulk optics and complex control mechanisms, which can be bulky, expensive, and power-intensive. Metamaterial-based optical components, such as waveguides, couplers, and splitters, offer a compact and efficient alternative for routing and distributing optical signals within the network. By engineering the electromagnetic properties of these components, it is possible to achieve low-loss signal routing, polarization control, and wavelength selectivity, enabling the implementation of reconfigurable and adaptive optical networks.

Deploying metamaterial-based components in fiber optic communication systems requires careful design, fabrication, and integration into existing infrastructure. Advanced fabrication techniques, such as nanoimprint lithography and electron beam lithography, are used to pattern metamaterial structures with subwavelength precision on optical substrates. These fabricated components are then integrated into fiber optic systems using standard assembly techniques, such as splicing, fusion bonding, or adhesive bonding, ensuring seamless compatibility with existing network architectures.

Moreover, numerical simulations and modeling tools play a crucial role in the design and optimization of metamaterial-based fiber optic components. Finite-difference time-domain (FDTD) simulations, rigorous coupled-wave analysis (RCWA), and finite element method (FEM) simulations are commonly used to predict the electromagnetic behavior of metamaterial structures and optimize their performance parameters. These simulation tools enable researchers and engineers to explore a wide range of design parameters and iterate rapidly to achieve the desired functionality.

In summary, metamaterials hold tremendous potential for enhancing fiber optic communication systems, offering unique capabilities for tailoring and controlling the propagation of light within optical fibers. By leveraging the extraordinary properties of metamaterials, researchers and engineers can develop innovative fiber optic components and devices with enhanced performance characteristics, enabling the realization of more efficient, reliable, and versatile communication networks. As research in metamaterials continues to advance, further breakthroughs are expected to drive the evolution of fiber optic technology and shape the future of telecommunications.

Chapter 7: Artificial Intelligence and Machine Learning in Fiber Optic Networks

Artificial Intelligence (AI) and Machine Learning (ML) have emerged as powerful tools in managing and optimizing fiber optic networks, offering innovative solutions to various challenges encountered in network operation and maintenance. These advanced technologies leverage data analytics, predictive modeling, and automation to enhance network performance, reliability, and efficiency. One of the key applications of AI and ML in fiber optic network management is fault detection and diagnosis, where these technologies analyze network data in real-time to identify and localize faults or anomalies that may degrade network performance. CLI commands can be utilized to deploy AI and ML-based fault detection systems, such as Cisco's IOS Embedded Event Manager (EEM) or Juniper's Junos Event Policies, which enable network administrators to define custom event detection and response mechanisms based on specific network conditions. These commands allow administrators to configure event-driven actions, such as generating alerts, executing scripts, or triggering remediation processes, to address detected faults automatically.

Furthermore, AI and ML algorithms can be deployed to predict and prevent network failures by analyzing historical performance data and identifying patterns indicative of potential issues. By proactively identifying and addressing emerging problems, network operators can minimize service disruptions and optimize network uptime. For instance, tools like IBM Watson for Networking and Arista's CloudVision Analytics utilize ML algorithms to analyze network telemetry data and predict impending failures or performance degradation, enabling preemptive maintenance actions to be taken before issues escalate. Deploying these tools typically involves configuring data collection mechanisms, such as streaming telemetry or SNMP polling, to gather relevant network performance metrics, and then applying ML models to analyze this data and generate actionable insights.

In addition to fault detection and prediction, AI and ML technologies play a crucial role in network optimization and resource allocation. These algorithms can analyze network traffic patterns, demand forecasts, and performance metrics to dynamically adjust routing, bandwidth allocation, and Quality of Service (QoS) parameters in real-time, optimizing network resource utilization and ensuring optimal service delivery. CLI commands such as Cisco's Quality of Service (QoS) configurations or Juniper's Class of Service (CoS) settings can be used to deploy AI and ML-driven network optimization policies, allowing administrators to prioritize critical traffic flows, allocate bandwidth dynamically, and enforce service level agreements (SLAs) based on real-time traffic conditions.

Moreover, AI and ML-based anomaly detection techniques can enhance network security by identifying suspicious or malicious activities that may indicate potential security threats, such as network intrusions or denial-of-service (DoS) attacks. By analyzing network traffic patterns and user behavior, these algorithms can detect deviations from normal network activity and trigger automated responses to mitigate security risks. For example, tools like Darktrace and Palo Alto Networks' Cortex XDR leverage ML algorithms to detect and respond to anomalous network behavior, such as unusual traffic patterns or unauthorized access attempts, by dynamically adjusting firewall policies, blocking suspicious traffic, or quarantining compromised devices. Deploying these security solutions involves configuring network monitoring sensors or agents to collect and analyze traffic data, as well as defining policies to respond to detected threats using CLI commands or graphical user interfaces (GUIs) provided by the security management platform.

Furthermore, AI and ML technologies can be leveraged to optimize network capacity planning and expansion by analyzing historical traffic data, demand forecasts, and performance metrics to predict future capacity requirements accurately. By identifying potential bottlenecks and capacity constraints, network operators can proactively upgrade network infrastructure and scale capacity to meet growing demand, minimizing the risk of service degradation or congestion. CLI commands such as Cisco's Traffic Engineering (TE) configurations or Juniper's Path Computation Element (PCE) settings can be used to deploy AI and ML-driven capacity planning solutions, allowing administrators to optimize network resource utilization, improve scalability, and reduce capital expenditures on network upgrades.

Additionally, AI and ML-based techniques can enhance network automation and self-healing capabilities, enabling autonomous network operation and reducing manual intervention. By analyzing network telemetry data and performance metrics in real-time, AI-driven automation platforms can detect and respond to network events or changes dynamically, such as traffic surges, link failures, or configuration drifts, by automatically adjusting network settings, reconfiguring routing paths, or triggering remediation actions. CLI commands such as Cisco's Embedded Event Manager (EEM) or Juniper's Event Script Engine (ESE) can be used to deploy AI and ML-driven automation workflows, allowing administrators to define custom event triggers and response actions based on specific network conditions.

In summary, the integration of AI and ML technologies into fiber optic network management systems offers significant advantages in terms of fault detection, performance optimization, security enhancement, and automation. By leveraging the power of data analytics, predictive modeling, and automation, network operators can improve network reliability, efficiency, and scalability while reducing operational costs and minimizing service disruptions. CLI commands play a crucial role in deploying AI and ML-driven network management solutions, enabling administrators to configure, monitor, and control network devices and services effectively. As AI and ML continue to advance, their applications in fiber optic network management are expected to evolve further, driving innovation and transformation in the telecommunications industry.

Predictive maintenance and optimization using AI techniques represent a pivotal advancement in fiber optic network management. With the ever-increasing complexity and scale of modern fiber optic infrastructures, traditional reactive maintenance approaches are no longer adequate. Instead, forward-thinking organizations are turning to AI-driven predictive maintenance strategies to enhance network reliability, reduce downtime, and optimize resource utilization.

One of the primary AI techniques leveraged in predictive maintenance is machine learning (ML). ML algorithms analyze vast amounts of historical data collected from fiber optic networks to identify patterns and trends indicative of potential failures or performance degradation. These algorithms can detect subtle anomalies that may go unnoticed by human operators, enabling early intervention before issues escalate into major disruptions.

Deploying ML for predictive maintenance typically involves several steps. Initially, relevant data sources such as network traffic, environmental conditions, and equipment telemetry are identified and integrated into a centralized data repository. This data is then preprocessed to ensure consistency and quality, which may involve cleaning, normalization, and feature engineering.

Once the data is prepared, ML models are trained using supervised, unsupervised, or semi-supervised learning techniques. Supervised learning involves training models on labeled data, where each sample is associated with a known outcome, such as equipment failure or degradation. Unsupervised learning, on the other hand, identifies patterns and clusters in unlabeled data, while semi-supervised learning combines elements of both approaches.

In the context of fiber optic networks, ML models can predict equipment failures, identify degradation in optical signals, and anticipate capacity constraints based on historical performance data. For example, recurrent neural networks (RNNs) and long short-term memory (LSTM) networks excel at time-series forecasting tasks, making them well-suited for predicting network outages or signal degradation.

CLI commands play a crucial role in deploying AI-driven predictive maintenance solutions in fiber optic networks. For instance, network administrators can use CLI commands to collect real-time performance metrics from network devices, such as optical power levels, signal-to-noise ratios, and error counts. These metrics are then fed into ML algorithms for analysis and model training.

Moreover, CLI commands facilitate the automation of routine maintenance tasks, such as device configuration changes or firmware updates, which can be orchestrated based on insights derived from predictive models. For instance, if a predictive model detects an impending hardware failure, CLI commands can automatically initiate a failover procedure to redirect traffic away from the affected equipment while maintenance is performed.

In addition to predictive maintenance, AI techniques are also instrumental in optimizing fiber optic network performance. By continuously analyzing network data and identifying opportunities for optimization, AI-driven systems can dynamically adjust network parameters to maximize throughput, minimize latency, and improve overall reliability.

Furthermore, AI-powered analytics can provide valuable insights into network usage patterns, enabling capacity planning and resource allocation decisions. For example, clustering algorithms can identify groups of users with similar traffic patterns, allowing network operators to tailor service offerings and allocate bandwidth more efficiently.

Overall, the integration of AI and ML techniques into fiber optic network management represents a paradigm shift towards proactive, data-driven maintenance strategies. By harnessing the power of AI to predict and prevent network failures, organizations can minimize downtime, reduce operational costs, and deliver a superior quality of service to end-users.

Chapter 8: Biophotonic Applications in Healthcare and Medicine

Biomedical imaging techniques utilizing biophotonics represent a cutting-edge domain at the intersection of photonics and biology, revolutionizing medical diagnostics and research. These techniques leverage the unique properties of light to visualize biological structures and processes at various scales, from the cellular level to whole organs, enabling non-invasive, high-resolution imaging with unparalleled precision and sensitivity.

One of the most widely used biophotonics imaging techniques is fluorescence imaging, which involves the excitation of fluorescent molecules within biological samples using specific wavelengths of light. CLI commands are often employed to control the illumination source and capture fluorescence signals. For instance, in a confocal fluorescence microscopy setup, the laser power and wavelength can be adjusted using CLI commands to optimize excitation efficiency and reduce photobleaching.

Another powerful biophotonics technique is optical coherence tomography (OCT), which enables high-resolution, cross-sectional imaging of biological tissues with micrometer-scale resolution. OCT CLI commands are used to control the scanning system and acquire depth-resolved images of tissue structures. For instance, in a swept-source OCT system, CLI commands can adjust the sweep rate and wavelength tuning range to optimize imaging depth and resolution.

Furthermore, multiphoton microscopy is a non-linear optical imaging technique that allows for deep-tissue imaging with subcellular resolution. CLI commands are utilized to control the laser scanning parameters and acquire multiphoton fluorescence signals. For instance, in a two-photon microscopy setup, CLI commands can adjust the laser power and scanning speed to optimize image contrast and signal-to-noise ratio.

In addition to fluorescence imaging and optical coherence tomography, other biophotonics techniques such as Raman spectroscopy, second-harmonic generation microscopy, and photoacoustic imaging are also widely used in biomedical research and clinical applications. CLI commands play a crucial role in configuring and controlling the instrumentation for these techniques, enabling researchers to acquire high-quality imaging data efficiently.

Moreover, advanced image processing algorithms are often applied to biophotonics imaging data to enhance contrast, remove noise, and extract quantitative information. CLI commands are used to execute these algorithms and analyze the resulting images. For instance, in a fluorescence microscopy workflow, CLI commands can apply deconvolution algorithms to improve image resolution and reveal fine structural details.

Biophotonics imaging techniques have a wide range of applications in biomedical research, clinical diagnostics, and therapeutic monitoring. For example, fluorescence imaging is used to study cellular dynamics, protein localization, and gene expression patterns in live cells and tissues. CLI commands can control the imaging parameters and automate image acquisition, allowing researchers to study dynamic biological processes over time.

Similarly, optical coherence tomography is employed in ophthalmology for imaging the retina and diagnosing retinal diseases such as macular degeneration and glaucoma. CLI commands enable clinicians to adjust the imaging parameters and capture high-resolution OCT scans of the retina, facilitating early detection and treatment of eye conditions.

Furthermore, multiphoton microscopy is used in neuroscience to study neuronal activity and synaptic connections in intact brain tissue. CLI commands can control the imaging system and acquire volumetric imaging data, enabling researchers to map neural circuits and investigate the neural basis of behavior and cognition.

Overall, biophotonics imaging techniques offer powerful tools for visualizing biological structures and processes with unprecedented detail and sensitivity. By leveraging CLI commands to control instrumentation and analyze imaging data, researchers and clinicians can advance our understanding of complex biological systems and develop innovative approaches for diagnosing and treating disease.

Therapeutic applications of biophotonics in medicine encompass a diverse array of techniques that utilize light-based technologies to diagnose, monitor, and treat various medical conditions. These innovative approaches harness the unique properties of light to deliver targeted therapies, improve treatment outcomes, and minimize patient discomfort and side effects.

Photodynamic therapy (PDT) is one such therapeutic modality that relies on the interaction between light, photosensitizing agents, and oxygen to selectively destroy cancer cells and other abnormal tissues. In PDT, a photosensitizer is administered to the patient and allowed to accumulate in the target tissue. Subsequently, light of a specific wavelength is delivered to the treatment site, activating the photosensitizer and generating cytotoxic reactive oxygen species that induce cell death. CLI commands can control the light source and monitor treatment parameters during PDT sessions, ensuring precise delivery of light energy to the target tissue while minimizing damage to surrounding healthy tissue.

Another emerging therapeutic application of biophotonics is photobiomodulation therapy (PBMT), also known as low-level laser therapy (LLLT). PBMT involves the use of low-power laser or light-emitting diode (LED) devices to stimulate cellular function and promote tissue repair and regeneration. CLI commands are utilized to adjust the laser or LED parameters, such as wavelength, power, and treatment duration, to optimize therapeutic outcomes. For example, in a PBMT session for wound healing, CLI commands can control the laser power and irradiation time to promote tissue healing and reduce inflammation.

Furthermore, optical coherence tomography (OCT) is increasingly being used in ophthalmology for the diagnosis and monitoring of various retinal diseases, including age-related macular degeneration (AMD), diabetic retinopathy, and glaucoma. OCT CLI commands enable clinicians to acquire high-resolution cross-sectional images of the retina and assess changes in retinal thickness, morphology, and microvasculature over time. For instance, in a patient with AMD, CLI commands can adjust the OCT scanning parameters to capture detailed images of drusen deposits and monitor disease progression.

Additionally, laser ablation therapy is a minimally invasive treatment modality that utilizes focused laser energy to precisely remove or ablate abnormal tissue while sparing adjacent healthy tissue. CLI commands are employed to control the laser parameters, such as wavelength, pulse duration, and energy fluence, to achieve the desired tissue effect. For example, in laser ablation therapy for the treatment of skin lesions, CLI commands can adjust the laser settings to selectively target and destroy the lesion while preserving the surrounding skin.

Moreover, fluorescence-guided surgery (FGS) is an innovative technique that uses fluorescent dyes or probes to enhance the visualization of tumor margins and guide surgical resection. CLI commands are utilized to control the fluorescence imaging system and adjust the excitation and emission wavelengths for optimal visualization of the fluorescent signal. For instance, in a fluorescence-guided resection of brain tumors, CLI commands can adjust the imaging parameters to differentiate between tumor tissue and normal brain tissue, helping surgeons achieve complete tumor removal while minimizing damage to healthy brain tissue.

Overall, therapeutic applications of biophotonics in medicine offer promising avenues for the diagnosis, monitoring, and treatment of various medical conditions. By leveraging CLI commands to control light-based therapies and imaging systems, clinicians can deliver targeted treatments with precision and efficacy, leading to improved patient outcomes and quality of life.

Chapter 9: Terahertz Communication Systems

Terahertz (THz) communication is an emerging technology that operates in the frequency range between microwave and infrared wavelengths, typically from 0.1 to 10 THz. This relatively unexplored portion of the electromagnetic spectrum offers promising opportunities for high-speed wireless communication, imaging, sensing, and spectroscopy applications. In THz communication systems, CLI commands play a crucial role in configuring and managing various components, such as THz sources, detectors, antennas, and signal processing units, to establish reliable and high-performance communication links.

One of the fundamental aspects of THz communication is the generation of THz signals, which can be achieved using different techniques such as photomixing, photonic-based approaches, and electronic-based methods. Photomixing involves the use of optical sources, such as lasers, to generate THz waves through nonlinear optical processes. CLI commands are used to control the laser parameters, including wavelength, power, and modulation, to produce stable and tunable THz signals. Additionally, photonic-based approaches utilize photonic devices, such as photodiodes and photomixers, to convert optical signals into THz radiation. CLI commands enable users to adjust the operating conditions of these devices for optimal THz generation efficiency. Furthermore, electronic-based methods employ electronic circuits, such as frequency multipliers and mixers, to directly generate THz signals from microwave sources. CLI commands are employed to configure the electronic components and set the desired frequency and power levels for THz signal generation.

Another key aspect of THz communication is the propagation of THz waves through various transmission media, including free space, waveguides, and fiber-optic cables. CLI commands are utilized to optimize the transmission parameters, such as frequency, modulation format, and polarization, to mitigate signal attenuation, dispersion, and interference effects. For instance, in free-space THz communication links, CLI commands can be used to adjust the alignment and orientation of the transmitter and receiver antennas to maximize signal strength and minimize multipath fading. Similarly, in waveguide-based systems, CLI commands enable users to configure the waveguide dimensions and material properties to achieve low-loss and high-speed transmission of THz signals. Moreover, in fiber-optic THz communication networks, CLI commands are employed to manage the optical amplifiers, dispersion compensators, and other components to maintain signal integrity and reliability over long-distance links.

Furthermore, modulation and coding techniques play a crucial role in THz communication systems to enhance spectral efficiency, data rate, and error performance. CLI commands are used to configure the modulators, demodulators, and error correction coding schemes to optimize the communication link performance. For example, in THz wireless communication systems, CLI commands can adjust the modulation format, such as amplitude modulation (AM), frequency modulation (FM), or phase modulation (PM), to achieve robust transmission in different channel conditions. Additionally, CLI commands are employed to set the coding rate, modulation order, and error correction coding scheme parameters to maximize data throughput and minimize bit error rate (BER) in THz communication links.

Moreover, antenna design and beamforming techniques are critical for achieving efficient THz signal transmission and reception in directional communication systems. CLI commands are utilized to configure the antenna arrays, antenna elements, and beamforming algorithms to optimize the spatial coverage, beamwidth, and sidelobe levels of the transmitted and received THz signals. For instance, in THz wireless communication networks, CLI commands can adjust the phase and amplitude weights of the antenna array elements to steer the beam towards the intended receiver and suppress interference from other directions. Additionally, in THz imaging and sensing applications, CLI commands enable users to control the scanning pattern and resolution of the antenna array to achieve high-resolution and real-time imaging of objects and surfaces.

Furthermore, signal processing and modulation techniques are essential for extracting information from THz signals and recovering the transmitted data in communication systems. CLI commands are employed to configure the signal processing algorithms, such as digital filtering, equalization, and synchronization, to mitigate channel impairments and noise effects in THz communication links. Additionally, CLI commands can be used to set the parameters of the demodulation and decoding algorithms to accurately recover the transmitted data from the received THz signals. For example, in THz wireless communication systems, CLI commands can adjust the adaptive equalizer coefficients and carrier frequency offset compensation to mitigate intersymbol interference (ISI) and frequency offset effects in multipath and Doppler-shifted channels.

In summary, the fundamentals of terahertz communication encompass various aspects, including THz signal generation, propagation, modulation, coding, antenna design, beamforming, signal processing, and modulation techniques. CLI commands play a crucial role in configuring and managing the different components and parameters of THz communication systems to achieve reliable, high-speed, and efficient wireless communication links. As THz communication technology continues to advance, CLI commands will remain essential for optimizing THz system performance and enabling a wide range of applications in wireless communication, imaging, sensing, and spectroscopy.

Terahertz (THz) communication systems hold immense promise for enabling ultra-fast wireless communication, high-resolution imaging, and sensing applications. However, they also face several significant challenges that must be addressed to realize their full potential.

One of the primary challenges in THz communication systems is the limited available spectrum. The THz frequency band is relatively underutilized compared to lower frequency bands, such as microwave and millimeter-wave bands. Consequently, there is a need for efficient spectrum management techniques to optimize spectrum utilization and mitigate interference in THz communication systems.

One approach to address the spectrum scarcity challenge is dynamic spectrum access (DSA), which allows THz communication systems to adaptively access and utilize available spectrum resources based on real-time demand and interference conditions. CLI commands play a crucial role in implementing DSA algorithms and protocols in THz communication systems. For example, the **spectrum-scan** command can be used to scan the THz spectrum and identify available frequency channels, while the **spectrum-select** command can be used to select the optimal channel for data transmission based on channel quality metrics such as signal-to-noise ratio (SNR) and interference level.

Another challenge in THz communication systems is the high path loss and atmospheric absorption at THz frequencies. THz waves are highly attenuated by atmospheric gases, water vapor, and particulate matter, limiting the range and coverage of THz communication links. To overcome this challenge, CLI commands can be used to optimize the transmission parameters, such as transmit power, modulation format, and antenna configuration, to maximize signal strength and coverage in THz communication systems. For instance, the **tx-**

power command can be used to adjust the transmit power of THz transmitters to compensate for path loss and atmospheric attenuation, while the **antenna-config** command can be used to configure the antenna beamwidth and gain for optimal coverage and link reliability.

Furthermore, another challenge in THz communication systems is signal dispersion and multipath fading, which can degrade signal quality and reliability, particularly in outdoor and indoor environments with complex propagation conditions. CLI commands can be employed to implement adaptive modulation and coding (AMC) techniques in THz communication systems to mitigate the effects of dispersion and fading. For example, the **modulation** and **coding** commands can be used to dynamically adjust the modulation scheme and coding rate based on channel conditions and link quality metrics, such as received signal strength and bit error rate (BER), to optimize data throughput and error performance in THz communication links.

Additionally, another challenge in THz communication systems is the lack of standardized protocols and interoperability among different THz devices and equipment. CLI commands can be utilized to configure and manage THz communication interfaces and protocols to ensure compatibility and interoperability between different THz devices and networks. For example, the **interface-config** command can be used to configure THz communication interfaces, such as modulation formats, frame formats, and synchronization mechanisms, to ensure seamless communication between THz devices from different vendors and manufacturers.

Despite these challenges, THz communication systems offer numerous opportunities for advancing wireless communication and enabling new applications in areas such as high-speed data transmission, wireless sensing, imaging, and spectroscopy. CLI commands play a vital role in realizing these opportunities by enabling efficient configuration, management, and optimization of THz communication systems. As THz communication technology continues to mature, addressing these challenges and leveraging these opportunities will be essential for unlocking the full potential of THz communication systems in various applications and industries.

Chapter 10: Beyond 5G: Next-Generation Fiber Optic Networks

The advent of 5G technology has sparked significant excitement and anticipation due to its potential to revolutionize wireless communication with unprecedented speed, capacity, and connectivity. However, despite its promises, 5G also comes with its set of limitations and challenges that need to be addressed for its successful deployment and widespread adoption.

One of the key features of 5G technology is its significantly higher data rates compared to previous generations of cellular networks. 5G promises to deliver peak data rates of up to 20 Gbps, enabling ultra-fast downloads, seamless streaming of high-definition content, and low-latency applications such as augmented reality (AR) and virtual reality (VR) experiences. However, achieving these high data rates requires the deployment of a dense network of small cells and millimeter-wave (mmWave) frequencies, which presents several challenges in terms of infrastructure deployment and signal propagation.

CLI commands are instrumental in configuring and optimizing 5G network parameters to achieve the desired data rates and network performance. For example, the **cell-config** command can be used to configure the parameters of small cells, such as transmit power, antenna tilt, and beamforming settings, to optimize coverage and capacity in densely populated areas. Similarly, the **spectrum-allocation** command can be used to allocate frequency bands for 5G networks, including both sub-6 GHz and mmWave bands, to maximize spectrum utilization and minimize interference.

Another key aspect of 5G technology is its ultra-low latency, which is essential for supporting real-time applications such as autonomous vehicles, remote surgery, and industrial automation. 5G promises to deliver latency as low as 1 millisecond (ms), significantly reducing the response time for critical applications and improving user experience. However, achieving ultra-low latency requires the deployment of edge computing resources and network slicing techniques to process data closer to the end-users and reduce the round-trip time for data transmission.

CLI commands play a crucial role in configuring and managing edge computing resources and network slices in 5G networks. For example, the **edge-config** command can be used to deploy edge computing nodes at strategic locations within the network, such as central offices or base stations, to minimize latency and optimize application performance. Similarly, the **network-slice** command can be used to create virtualized network slices with dedicated resources and quality-of-service (QoS) parameters to meet the requirements of specific applications, such as low-latency, high-throughput, or ultra-reliability.

Despite its potential, 5G technology also faces several limitations and challenges that need to be addressed for its widespread deployment and adoption. One of the main limitations of 5G is its limited coverage and penetration in indoor and rural areas. The high frequencies used in mmWave bands have limited range and are susceptible to blockage by obstacles such as buildings and trees, making it challenging to provide seamless coverage in urban environments and rural areas.

To address these coverage limitations, CLI commands can be used to optimize the deployment of 5G small cells and repeaters to improve coverage and penetration in indoor and rural areas. For example, the **small-cell-deploy** command can be used to identify suitable locations for deploying small cells based on coverage maps and network capacity requirements. Similarly, the **repeater-config** command can be used to configure repeater parameters, such as transmit power and antenna orientation, to extend coverage and improve signal strength in areas with poor coverage.

Another limitation of 5G technology is its susceptibility to interference and signal degradation due to factors such as weather conditions, atmospheric attenuation, and co-channel interference from neighboring cells. CLI commands can be used to mitigate interference and optimize signal quality in 5G networks by adjusting transmission parameters and frequency allocations. For example, the **interference-**

management command can be used to detect and mitigate interference sources in the network, such as adjacent-channel interference or signal reflections, to improve network performance and reliability.

Furthermore, another limitation of 5G technology is its high energy consumption and power requirements, particularly in dense urban areas with a large number of small cells and base stations. CLI commands can be used to optimize power management and energy efficiency in 5G networks by implementing techniques such as sleep mode, dynamic power scaling, and energy harvesting. For example, the **power-management** command can be used to adjust the power-saving settings of small cells and base stations based on traffic patterns and network load to minimize energy consumption while maintaining network performance.

In summary, while 5G technology offers significant advancements in terms of speed, capacity, and connectivity, it also comes with its set of limitations and challenges that need to be addressed for its successful deployment and widespread adoption. CLI commands play a crucial role in configuring, optimizing, and managing 5G networks to overcome these limitations and deliver the promised benefits of ultra-fast, low-latency wireless communication.

As the demand for high-speed, reliable, and scalable communication networks continues to grow, researchers and industry experts are exploring various future directions and technologies to enhance the performance and capabilities of next-generation fiber optic networks. These advancements aim to address the increasing bandwidth requirements, reduce latency, improve energy efficiency, and enable new applications and services that will shape the future of telecommunications.

One of the key areas of focus for future fiber optic networks is the development of advanced modulation and multiplexing techniques to increase the transmission capacity and spectral efficiency of optical fibers. Current fiber optic systems primarily rely on wavelength-division multiplexing (WDM) to transmit multiple channels of data simultaneously over a single fiber by using different wavelengths of light. However, researchers are exploring new multiplexing schemes such as spatial-division multiplexing (SDM), mode-division multiplexing (MDM), and frequency-division multiplexing (FDM) to further increase the data rates and capacity of optical fibers.

CLI commands are essential for configuring and optimizing advanced modulation and multiplexing techniques in fiber optic networks. For example, the **sdm-config** command can be used to configure the parameters of spatial-division multiplexing systems, such as the number of spatial modes supported and the coupling efficiency between modes. Similarly, the **mdm-config** command can be used to configure mode-division multiplexing systems, such as the mode coupling coefficients and mode excitation conditions, to maximize the transmission capacity and spectral efficiency of optical fibers.

Another emerging technology that holds great promise for future fiber optic networks is quantum communication, which leverages the principles of quantum mechanics to enable secure and unbreakable communication channels. Quantum communication techniques such as quantum key distribution (QKD) and quantum teleportation offer unprecedented levels of security by using quantum states of light to encode and transmit information. These techniques are immune to eavesdropping and interception, making them ideal for securing sensitive data and communications.

CLI commands play a crucial role in configuring and managing quantum communication systems in fiber optic networks. For example, the **qkd-config** command can be used to configure the parameters of quantum key distribution systems, such as the quantum states used for encoding and decoding information, the error correction codes applied to the transmitted data, and the authentication protocols used to verify the integrity of the communication channel. Similarly, the **teleportation-config** command can be used to configure quantum teleportation systems, such as the entangled photon sources, the quantum measurement devices, and the quantum channel parameters.

In addition to advanced modulation and multiplexing techniques and quantum communication, other future directions for fiber optic networks include the development of new materials and components, such as metamaterials, photonic crystals, and plasmonic devices, to enable novel functionalities and applications. These materials and devices can be used to manipulate and control the propagation of light in optical fibers, enabling new capabilities such as on-chip integration, nonlinear signal processing, and enhanced sensing and imaging.

CLI commands are indispensable for configuring and optimizing new materials and components in fiber optic networks. For example, the **metamaterials-config** command can be used to configure metamaterial-based devices, such as metasurfaces and metamaterial lenses, to control the phase, amplitude, and polarization of light for various applications, such as beam steering, focusing, and polarization manipulation. Similarly, the **photonic-crystals-config** command can be used to configure photonic crystal structures, such as photonic bandgaps and defect modes, to control the propagation of light in optical fibers for applications such as wavelength filtering, dispersion compensation, and mode conversion.

Furthermore, advancements in fiber optic network management and control are essential for enabling the efficient operation and maintenance of future networks. CLI commands are crucial for configuring and monitoring network resources, analyzing network performance, and diagnosing and troubleshooting network issues. For example, the **network-config** command can be used to configure the parameters of network elements such as routers, switches, and optical amplifiers, while the **network-monitor** command can be used to monitor network traffic, latency, and packet loss in real-time.

In summary, future fiber optic networks will continue to evolve and innovate to meet the growing demands for high-speed, reliable, and secure communication. Advanced modulation and multiplexing techniques, quantum communication, new materials and components, and network management and control are among the key technologies that will shape the future of fiber optic networks. CLI commands will play a critical role in deploying, configuring, and optimizing these technologies to realize the full potential of next-generation fiber optic networks.

Conclusion

In summary, the "Fiber Optics: Networking and Data Transmission in Action" book bundle provides a comprehensive overview of fiber optics technology, covering everything from the fundamentals to advanced techniques and emerging trends.

Book 1, "Fiber Optics 101: A Beginner's Guide to Networking and Data Transmission," serves as an excellent starting point for individuals new to the field, offering clear explanations of basic concepts and principles.

Book 2, "Mastering Fiber Optic Networks: Advanced Techniques and Applications," delves deeper into the subject, exploring advanced networking techniques, troubleshooting methods, and practical applications in various industries.

Book 3, "Fiber Optic Infrastructure Design and Implementation: Practical Strategies for Professionals," offers practical guidance for designing, deploying, and managing fiber optic networks, providing professionals with the knowledge and skills needed to tackle real-world projects.

Book 4, "Cutting-Edge Fiber Optics: Emerging Technologies and Future Trends in Networking," explores the latest innovations and future directions in fiber optics technology, offering insights into upcoming trends, such as quantum communication, terahertz transmission, and photonic integrated circuits.

Collectively, these books offer a comprehensive and insightful look into the world of fiber optics, making them essential resources for anyone involved in networking and data transmission. Whether you are a beginner looking to learn the basics or a seasoned professional seeking to stay updated on the latest advancements, this book bundle has something to offer for everyone.